Principles of Enzyme Kinetics

Principles of Enzyme Kinetics

ATHEL CORNISH-BOWDEN M.A., D.Phil. (Oxon)

Lecturer in Biochemistry, University of Birmingham

BUTTERWORTHS
LONDON - BOSTON
Sydney - Wellington - Durban - Toronto

THE BUTTERWORTH GROUP

ENGLAND

Butterworth & Co (Publishers) Ltd
London: 88 Kingsway, WC2B 6AB

AUSTRALIA

Butterworths Pty Ltd
Sydney: 586 Pacific Highway, NSW 2067
    Also at Melbourne, Brisbane
    Adelaide and Perth

SOUTH AFRICA

Butterworth & Co (South Africa) (Pty) Ltd
Durban: 152–154 Gale Street

NEW ZEALAND

Butterworths of New Zealand Ltd
Wellington: 26–28 Waring Taylor Street, 1

CANADA

Butterworth & Co (Canada) Ltd
Toronto: 2265 Midland Avenue,
    Scarborough, Ontario, M1P 4S1

USA

Butterworths (Publishers) Inc
161 Ash Street,
    Reading, Boston, Mass. 01867

First published 1976

ISBN 0 408 70721 6

© Butterworth & Co 1976

Library of Congress Cataloging in Publication Data

Cornish-Bowden, Athel.
    Principles of enzyme kinetics.

    Includes bibliographical references and index.
    1. Enzymes. I. Title. [DNLM: 1. Enzymes.
2. Kinetics. QU135 C818p]
QP601.C757   574.1'925   75-33717
ISBN 0-408-70721-6

Typeset in 10/11pt Monophoto Times New Roman
Printed in England
by Butler & Tanner Ltd
London & Frome

Foreword

Kinetics is a subject of mystery and power. It is mysterious to a great many scientists, because at some point in their careers they acquired a fear and awe of mathematical techniques. It is powerful to many others because no other tool in all of science has such universality. Although this potential for mystery and power is present in many techniques, kinetics seems to be special in its capacity for polarizing individuals.

Those who do not understand and fear mathematics tend either to have excessive admiration for kinetics ('It's too difficult for me') or excessive contempt ('Kinetics can never prove a mechanism; it can merely disprove one'). The latter statement, while true, applies to any scientific technique, so one might ask why kinetics is a special target for criticism. The answer would seem to lie in the twin pitfalls that (a) the explanations of kinetic procedures are frequently confusing and imprecise and (b) the power of kinetics is frequently over-stated and therefore leads to error.

In this book, Athel Cornish-Bowden does an admirable job in steering between these Scylla and Charybdis of kinetics. Firstly, the core of enzyme kinetics is explained in a simple manner which the serious biologist, who may not be 'a mathematical type', can follow readily. Secondly, the limitations of the technique and the dangers of excessive extrapolation are clearly outlined. The reader is given a powerful weapon and warned that it can backfire if not handled properly. This is a well disciplined book. It does not contain everything that is known about kinetics; and that is one of its virtues. It has distilled some of the most important areas of kinetics, treating illustrative sections with rigour and clarity, which should help to provide more enthusiasm for kinetics and more recognition of its power for applications in biology and enzymology.

DANIEL E. KOSHLAND, JR.

Contents

Preface

This book is written primarily for first-year research students in enzyme kinetics, but I hope that it will also prove useful to more advanced research workers and to final-year undergraduates. For the student beginning research, particularly one with a first degree in chemistry or biology, it is often difficult to find a text that goes beyond an elementary and idealized account of enzyme kinetics, but does not assume a large amount of specialized background knowledge and understanding. There are several topics in enzyme kinetics, such as the derivation of steady-state rate equations, the analysis of progress curves and the statistical treatment of results, that are important enough to be covered at an elementary level but are usually discussed inadequately or not at all.

With a proper understanding of the principles of enzyme kinetics, the whole subject comes within reach. It becomes more complicated, although not more difficult, as it is developed. For this reason, it is important to cover the elementary aspects thoroughly. If I have erred, therefore, I hope that it is in the direction of over-explaining the simple, rather than omitting to explain the difficult. However, I have not tried to write an exhaustive treatise: there is little mention of three-substrate reactions, for example, not because they are not important, but because they can be studied within the framework developed for the study of simpler reactions. In short, they contribute complexity rather than understanding.

Although I hope I have been consistent in important matters in this book, I have deliberately avoided any attempt to use a slavishly consistent system of nomenclature and symbolism. There are circumstances in which S is an appropriate symbol for substrate, for example, and others in which A, B . . ., are preferable; I have therefore used both. Similarly, it is an unfortunate fact that one of the two principal theories of co-operativity has been defined in terms of dissociation constants and the other in terms of association constants, but it would be a hindrance rather than a help to the student to re-define them in a consistent way, because this would confuse any attempt to read the original literature.

The emphasis throughout this book is on understanding enzyme kinetics, and not on information about specific enzymes. It is in no sense a catalogue of the properties of enzymes. Not only are there already several books that fulfil that role admirably, but there also seems to be a real need for a book that discusses the principles of enzyme kinetics at an intermediate level. I hope that this book will help to fill that need.

Many mathematically inclined books begin with a descriptive introduction, designed, presumably, to lull the reader into a false sense of security. This book does not follow that format, because I believe that it should be made clear at the outset that enzyme kinetics is not a subject for anyone who is frightened of simple algebra or simple calculus. Chapter 1 is a résumé of chemical kinetics, much of which should be familiar to the reader, but it is included in order to establish the knowledge that will be assumed in later chapters. It also includes a brief discussion of dimensional analysis, which I believe to be by far the most powerful simple method for detecting algebraic errors.

Chapters 2, 4, 5 and 6 cover the essential characteristics of steady-state kinetics as taught in innumerable biochemistry courses, and require little special discussion here. I have deviated slightly from common practice by treating V/K_m as a parameter in its own right, at least as important as K_m, because many aspects of enzyme kinetics are far simpler to understand and classify in terms of V and V/K_m rather than V and K_m. Some may feel that the section in Chapter 6 on temperature dependence is rather short. This is because I feel that very few of the large number of studies on the temperature dependence of enzyme activity are of any value. It is rare for conditions to be sufficiently favourable that a temperature study can usefully be carried out.

Chapter 3 sets out to explain as simply as possible the most useful methods for deriving steady-state rate equations: the student of complex mechanisms soon discovers that the method taught in the context of very simple mechanisms is virtually useless because of the hopelessly complicated algebra that it engenders. Although the King–Altman method has been outlined in several textbooks, its principle is to be found only, as far as I know, in the original, difficult, paper. However, I believe that anyone who often uses such an important method ought to have some understanding of its theoretical basis, and so I have tried to explain this as simply as possible. The chapter also includes some important developments from the King–Altman method that have been made in recent years.

The study of co-operativity (Chapter 7) has developed apart from the mainstream of enzyme kinetics, and it has often been neglected in textbooks. It has developed with its own conventions, such as the more common use of association constants than dissociation constants. In this chapter particularly, and to some extent throughout the book, the temptation to invent new symbols and terminology has been strong, but I have not consciously succumbed to it anywhere, except for the use of h for the Hill coefficient. This exception seemed justified by the very strong objection to n, which is grossly misleading, and the typographically cumbersome nature of n_H (particularly when used as an exponent).

Chapter 8 concerns an aspect of enzyme kinetics that has been almost completely ignored by biochemists for 60 years, for reasons that have lost much of their original force. I believe that it is time for integrated rate equations to regain the respectability that they lost with the classic work of Michaelis and Menten.

Chapter 9 is an introduction to the study of fast reactions, but it does not pretend to be a comprehensive account, which would require a separate book. Instead, I have tried to cover those aspects of fast reactions that ought to be familiar to anyone who is working mainly on studies in the steady state, but who feels that the need for transient-state studies may arise occasionally.

Chapter 10 is an introduction to the statistical aspects of enzyme kinetics. Many biochemists apparently believe that it is unnecessary to understand this topic, but they deceive themselves. The continued widespread use of the Lineweaver–Burk plot is evidence of the laziness of the majority who cannot be bothered to discover the most basic information about data analysis.

I am grateful to Dr. J. R. Knowles and to Dr. D. E. Koshland for stimulating and developing my interest in enzyme kinetics, and to several colleagues, particularly Dr. R. Eisenthal, Mr. A. C. Storer, Dr. C. W. Wharton and Dr. E. A. Wren, for many helpful comments on the first draft of this book. It has gained much from their advice, and has lost numerous errors. Doubtless some remain, as all books that contain many equations contain errors, and I shall greatly appreciate it if they can be brought to my attention.

ATHEL CORNISH-BOWDEN

1

Basic Principles of Chemical Kinetics

1.1 Order of reaction

A chemical reaction can be classified either according to its *molecularity* or according to its *order*. The molecularity is defined by the number of molecules that are altered in the reaction. Thus, a reaction A → products is *unimolecular* or *monomolecular*, a reaction A + B → products or 2A → products is *bimolecular*, and a reaction A + B + C → products is *trimolecular* or *termolecular*. (The illogical variations in prefixes is a consequence of the unfortunate propensity of scientists for inventing new words from what they imagine to be classical roots without first determining what the roots mean. In this book, I shall use the first of each of the above alternatives, albeit with misgivings.) The order is a description of the number of concentration terms multiplied together in the rate equation. Hence, in a first-order reaction, the rate is proportional to one concentration, in a second-order reaction it is proportional to two concentrations or to the square of one concentration, and so on.

For a simple reaction that consists of a single step, or for each step in a complex reaction, the order is generally the same as the molecularity. However, many reactions consist of a sequence of unimolecular and bimolecular steps, and the molecularity of the complete reaction need not be the same as its order. Reactions of molecularity greater than 2 are common, but reactions of order greater than 2 are very rare. It should also be noted that neither the molecularity nor the order of a reverse reaction need be the same as the corresponding molecularity or order of the forward reaction. This is an important consideration for metabolic reactions, which are often reversible and can be made to proceed in either direction by adjusting the concentrations of reactants.

For a first-order reaction A → P, the velocity, v, can be expressed as follows:

$$v = \frac{\mathrm{d}p}{\mathrm{d}t} = ka = k(a_0 - p) \tag{1.1}$$

1

where a and p are the concentrations of A and P, respectively, and are related by the equation $a + p = a_0$, t is time and k is a *first-order rate constant*. This equation can readily be integrated as follows:

$$\int \frac{dp}{a_0 - p} = \int k\, dt$$

Therefore,

$$-\ln(a_0 - p) = kt + \alpha$$

where α is a constant of integration, which can be evaluated by defining the time scale so that $a = a_0$, $p = 0$ when $t = 0$. Then, $\alpha = -\ln(a_0)$, and so

$$\ln[(a_0 - p)/a_0] = -kt$$

which can be rearranged to give

$$p = a_0[1 - \exp(-kt)] \tag{1.2}$$

It is important to note that the constant of integration, α, was included in this derivation, evaluated and found to be non-zero. Constants of integration must always be included and calculated when kinetic equations are integrated; they are very rarely found to be zero.

A simple bimolecular reaction, $2A \rightarrow P$, is likely to be second order, with rate v given by

$$v = dp/dt = ka^2 = k(a_0 - 2p)^2 \tag{1.3}$$

where k is now a *second-order rate constant*. (Notice that conventional symbolism does not, unfortunately, indicate the order of a rate constant.) Then,

$$\int \frac{dp}{(a_0 - 2p)^2} = \int k\, dt$$

Therefore,

$$\frac{1}{2(a_0 - 2p)} = kt + \alpha$$

and, putting $p = 0$ when $t = 0$, we have $\alpha = 1/2a_0$, so that

$$p = \frac{a_0^2 kt}{1 + 2a_0 kt} \tag{1.4}$$

Reactions of this type are not unknown, but they are rare, and bimolecular reactions are much more commonly of the type $A + B \rightarrow P$, in which the two reacting molecules are different:

$$v = dp/dt = kab = k(a_0 - p)(b_0 - p) \tag{1.5}$$

In this instance, we have

$$\int dp/[(a_0 - p)(b_0 - p)] = \int k\, dt$$

which can be integrated to give

2

$$\left(\frac{1}{b_0-a_0}\right)\ln\left(\frac{b_0-p}{a_0-p}\right) = kt+\alpha$$

and putting $p = 0$ when $t = 0$ and rearranging, we obtain

$$\ln\left[\frac{a_0(b_0-p)}{b_0(a_0-p)}\right] = (b_0-a_0)kt$$

or

$$\frac{a_0(b_0-p)}{b_0(a_0-p)} = \exp[(b_0-a_0)kt] \tag{1.6}$$

The following special case of this result is of interest: if $a_0 \gg b_0$, then p must be insignificant compared with a_0 at every stage in the reaction, and so $(a_0 - p)$ can be written simply as a_0. In this case, equation 1.6 simplifies to

$$p = b_0[1 - \exp(-ka_0 t)]$$

which is of exactly the same form as equation 1.2, the equation for a first-order reaction. This type of reaction is known as a *pseudo-first-order* reaction, and ka_0 is a *pseudo-first-order rate constant*. The situation arises most often when one of the reactants is the solvent, as in most hydrolysis reactions, but it is also advantageous to set up pseudo-first-order conditions deliberately, in order to simplify evaluation of the rate constant, as we shall discuss in Section 1.5.

Trimolecular reactions, such as $A + B + C \rightarrow P$, do not usually consist of a single trimolecular step, and consequently they are not usually third order. Instead, the reaction usually consists of two or more *elementary steps*, such as

$$A + B \rightarrow X$$
$$X + C \rightarrow P$$

If one step in such a reaction is much slower than the others, the rate of the complete reaction is equal to the rate of the slow step, which is accordingly known as the *rate-determining* (or *rate-limiting*) step. If there is no clearly defined rate-determining step, the rate equation is likely to be complex and to have no constant order. Some trimolecular reactions do display third-order kinetics, with $v = kabc$, where k is now a *third-order rate constant*, but it is *not* necessary to assume a three-body collision (which is inherently very unlikely) in order to account for this. Instead, we can assume a two-step mechanism, as above but with the first step rapidly reversible, so that the concentration of X is given by $x = Kab$, where K is an equilibrium constant. The rate of the reaction is determined by that of the slow second step:

$$v = k'xc = k'Kabc$$

where k' is the rate constant for the second step. Hence the observed third-order rate constant is actually the product of a second-order rate constant and an equilibrium constant.

Some reactions are observed to be *zero order*, i.e. the rate appears to be constant, independent of the concentration of reactant. If a reaction is zero order with respect to only one reactant, this may simply mean that the re-

3

actant enters the reaction after the rate-determining step. However, some reactions are zero order overall, i.e. independent of all reactant concentrations. Such reactions are invariably catalysed reactions and occur if every reactant is present in such large excess that the full potential of the catalyst is realized. Examples of this behaviour will be seen when enzyme catalysis is discussed.

1.2 Determination of the order of a reaction

The simplest means of determining the order of a reaction is to determine the rate at different concentrations of the reactants. Then a plot of log(rate) against log(concentration) gives a straight line with a slope equal to the order. If all of the reactant concentrations are altered in a constant ratio, the slope of the line is the overall order. It is usually useful to know the order with respect to each reactant, however, which can be found by altering the concentration of each reactant separately, keeping the other concentrations constant. Then the slope of the line will be equal to the order with respect to the variable reactant. For example, if the rate is second order in A and first order in B,

$$\mathrm{d}p/\mathrm{d}t = ka^2b$$

then

$$\log \mathrm{d}p/\mathrm{d}t = \log k + 2 \log a + \log b$$

Hence a plot of $\log \mathrm{d}p/\mathrm{d}t$ against $\log a$ (with b held constant) will have a slope of 2, and a plot of $\log \mathrm{d}p/\mathrm{d}t$ against $\log b$ (with a held constant) will have a slope of 1. Both plots should give the same intercept of $\log k$ on the $\log v$ axis, so they provide a useful check on one another. It is important to realize that if the rates are determined from the slopes of the progress curve (i.e. a plot of concentration against time), the concentrations of all of the reactants will change. Therefore, if valid results are to be obtained, either the initial concentrations of all the reactants must be in stoichiometric ratio, in which event the overall order will be found, or (more usually) the 'constant' reactants must be in large excess at the start of the reaction, so that the changes in their concentrations are insignificant. If neither of these alternatives is possible or convenient, the rates must be obtained from a set of measurements of the slope at zero time, i.e. of initial rates. This method is usually preferable for kinetic measurements of enzyme-catalysed reactions, because the progress curves of enzyme-catalysed reactions often do not rigorously obey the simple rate equations for extended periods of time. In practice, the progress curve for an enzyme-catalysed reaction often requires a more complicated equation than the integrated form of the rate equation derived for the initial rate.

1.3 Dimensions of rate constants

Dimensional analysis is a technique that deserves to be used much more

4

widely than it is. Concentrations can be expressed in M (or mol l^{-1}), and reaction rates in $M\,s^{-1}$. In an expression such as $v = ka$, therefore, the rate constant k must be expressed in s^{-1} if the left- and right-hand sides of the equation are to have the same dimensions. All first-order rate constants have the dimension $(\text{time})^{-1}$, and by a similar argument second-order rate constants have the dimensions $(\text{concentration})^{-1}(\text{time})^{-1}$, third-order rate constants $(\text{concentration})^{-2}(\text{time})^{-1}$ and zero-order rate constants $(\text{concentration})(\text{time})^{-1}$.

Knowledge of the dimensions of rate constants provides a useful method of checking the correctness of derived equations: the left- and right-hand sides of an equation (or inequality) must always have the same dimensions, and all of the terms in a summation must have the same dimensions. For example, if $(1+t)$ occurs in an equation, where t has the dimension (time), then either the equation is incorrect, or the 1 is a time that happens to have a numerical value of 1 unit. Quantities with different dimensions can be multiplied or divided, but must not be added or substracted. Thus, if k_1 is a first-order rate constant and k_2 is a second-order rate constant, a statement such as $k_1 \gg k_2$ is meaningless, just as $5\,g \gg 25\,°C$ is meaningless. However, a quantity such as $k_2 a$ has the dimensions $(\text{concentration})^{-1}(\text{time})^{-1}(\text{concentration})$, i.e. $(\text{time})^{-1}$, and thus has the dimensions of a first-order rate constant, and *can* be compared with other first-order constants.

This discussion may seem to be obvious, but it is surprising how often improper comparisons between rate constants are made, leading to erroneous conclusions. Rate constants and equilibrium constants are particularly liable to this type of error, because the algebraic symbols used for them do not usually provide any indication of the dimensions, and almost identical symbols are used for different types of constant.

1.4 Reversible reactions

Many reactions are readily reversible, and the back reaction must be allowed for in the rate equation:

$$A \underset{k_{-1}}{\overset{k_{+1}}{\rightleftarrows}} P$$

$$a_0 - p \qquad p$$

In this case,

$$v = dp/dt = k_{+1}(a_0 - p) - k_{-1}p = k_{+1}a_0 - (k_{+1} + k_{-1})p \qquad (1.7)$$

Therefore,

$$\int \frac{dp}{k_{+1}a_0 - (k_{+1} + k_{-1})p} = \int dt$$

and

$$\frac{\ln[k_{+1}a_0 - (k_{+1} + k_{-1})p]}{-(k_{+1} + k_{-1})} = t + \alpha$$

5

Setting $p = 0$ when $t = 0$ gives $\alpha = -\ln(k_{+1}a_0)/(k_{+1}+k_{-1})$ and so

$$\ln\left[\frac{k_{+1}a_0-(k_{+1}+k_{-1})p}{k_{+1}a_0}\right] = -(k_{+1}+k_{-1})t$$

Therefore,

$$p = \frac{k_{+1}a_0\{1-\exp[-(k_{+1}+k_{-1})t]\}}{k_{+1}+k_{-1}} \qquad (1.8)$$

A slightly more complex example is provided by a reversible bimolecular reaction:

$$A+B \underset{k_{-1}}{\overset{k_{+1}}{\rightleftharpoons}} P$$

In this case,

$$v = dp/dt = k_{+1}ab - k_{-1}p \qquad (1.9)$$

After infinite time, the net rate becomes zero, and the reactants reach their equilibrium concentrations, a_∞, b_∞ and p_∞, i.e.

$$k_{+1}a_\infty b_\infty - k_{-1}p_\infty = 0$$

and so

$$k_{-1} = k_{+1}a_\infty b_\infty/p_\infty$$

and

$$v = dp/dt = k_{+1}ab - (k_{+1}a_\infty b_\infty p/p_\infty)$$

Now, by the stoichiometry of the reaction, $a = a_0 - p$ and $b = b_0 - p$, and so

$$dp/dt = k_{+1}(a_0-p)(b_0-p) - [k_{+1}(a_0-p_\infty)(b_0-p_\infty)p/p_\infty]$$
$$= (k_{+1}/p_\infty)(a_0b_0 - pp_\infty)(p_\infty - p)$$

Therefore,

$$\int \frac{dp}{(a_0b_0-pp_\infty)(p_\infty-p)} = \int \frac{k_{+1}\,dt}{p_\infty}$$

Integration of this equation gives

$$\left(\frac{1}{-a_0b_0+p_\infty^2}\right)\ln\left(\frac{p_\infty-p}{a_0b_0-pp_\infty}\right) = \frac{k_{+1}t}{p_\infty} + \alpha$$

Putting $p = 0$ when $t = 0$ gives $\alpha = \left(\dfrac{1}{-a_0b_0+p_\infty^2}\right)\ln\left(\dfrac{p_\infty}{a_0b_0}\right)$

Therefore,

$$\left(\frac{1}{-a_0b_0+p_\infty^2}\right)\ln\left[\frac{a_0b_0(p_\infty-p)}{p_\infty(a_0b_0-pp_\infty)}\right] = \frac{k_{+1}t}{p_\infty} \qquad (1.10)$$

This equation can be rearranged to give a (rather complicated) expression for p, if desired.

The differential equations that result from consideration of reactions that involve sequences of steps including reversible steps, such as

$$A + B \rightleftharpoons X \rightarrow P$$

cannot in general be solved explicitly. Numerical solutions can be obtained with the help of modern computing methods, but as uncatalysed reactions of this type are not of great importance in biochemistry, they will not be considered further. Catalysed reactions of this type will be considered later.

1.5 Determination of first-order rate constants

Very many reactions are first-order in each reactant, and in these cases it is often possible to carry out the reaction under pseudo-first-order conditions overall by keeping every reactant except one in large excess. Thus, in many practical situations, the problem of determining rate constants can be reduced to the problem of determining the rate constant for a first-order reaction. We have seen (equation 1.2) that for a simple first-order reaction,

$$p = a_0[1 - \exp(-kt)]$$

and in the more general reversible case (equation 1.8),

$$p = \left(\frac{k_{+1}a_0}{k_{+1}+k_{-1}} \right)\{1 - \exp[-(k_{+1}+k_{-1})t]\}$$

Now, $k_{+1}a_0/(k_{+1}+k_{-1}) = p_\infty$, the equilibrium value of p, as the exponential term vanishes when t is large, and so

$$p_\infty - p = p_\infty \exp[-(k_{+1}+k_{-1})t] \tag{1.11}$$

Therefore,

$$\ln(p_\infty - p) = \ln p_\infty - (k_{+1}+k_{-1})t$$

or, more conveniently,

$$\log(p_\infty - p) = \log p_\infty - [(k_{+1}+k_{-1})t/2.303]$$

Thus, a plot of $\log(p_\infty - p)$ against t gives a straight line of slope $-(k_{+1}+k_{-1})/2.303$.

Guggenheim (1926) pointed out a major objection to this plot, in that it depends very heavily on an accurate value of p_∞. In the general case where $p_\infty \neq a_0$, an accurate value of p_∞ is difficult to obtain, and even in the special case of an irreversible reaction when $p_\infty = a_0$, the instantaneous concentration of A at zero time may be difficult to measure accurately. Guggenheim suggested measuring two sets of values, p_i and p_i', at times t_i and t_i', such that every $t_i' = t_i + \tau$, where τ is a constant. Then, from equation 1.11,

$$\left. \begin{array}{l} p_\infty - p_i = p_\infty \exp[-(k_{+1}+k_{-1})t_i] \\ p_\infty - p_i' = p_\infty \exp[-(k_{+1}+k_{-1})(t_i+\tau)] \end{array} \right\} \tag{1.12}$$

By subtraction,

$$p'_i - p_i = p_\infty\{1 - \exp[-(k_{+1}+k_{-1})\tau]\}\exp[-(k_{+1}+k_{-1})t_i]$$

Therefore,

$$\ln(p'_i - p_i) + (k_{+1}+k_{-1})t_i = \text{constant} \tag{1.13}$$

So, a plot of $\ln(p'_i - p_i)$ against t_i gives a straight line of slope $-(k_{+1}+k_{-1})$. This is known as a *Guggenheim plot*, and it has the major advantage that it does not require an estimate of p_∞. As k_{+1}/k_{-1} is equal to the equilibrium constant, which can be estimated independently, the values of the individual rate constants k_{+1} and k_{-1} can be calculated from the two combinations.

Equations 1.12 can alternatively be combined by dividing one by the other, to give

$$\frac{p_\infty - p_i}{p_\infty - p'_i} = \exp[(k_{+1}+k_{-1})\tau]$$

which can be rearranged to give

$$p'_i = p_\infty\{1 - \exp[-(k_{+1}+k_{-1})\tau]\} + p_i \exp[-(k_{+1}+k_{-1})\tau] \tag{1.14}$$

Thus, a plot of p'_i against p_i also gives a straight line, of slope $\exp[-(k_{+1}+k_{-1})\tau]$. This plot was suggested by Kézdy, Jaz and Bruylants (1958) and by Swinbourne (1960). Its accuracy is about the same as that of the more widely used Guggenheim plot, and it is easier to plot. As $p'_i = p_i = p_\infty$ when $t \to \infty$, this plot provides a simple method of estimating p_∞: if the line $p'_i = p_i$ is drawn, the point at which it intersects the (extrapolated) plot of p'_i against p_i gives the value of p_∞. Both types of plot are illustrated in *Figure 1.1*.

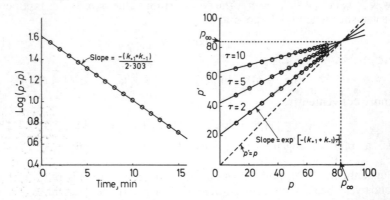

Figure 1.1 Determination of first-order rate constants by means of (left) Guggenheim plot and (right) Kézdy–Swinbourne plot: In both cases p *and* p' *are the concentrations of product at times* τ *apart, with* τ = 5 *min for the Guggenheim plot and* τ = 2, 5 *or* 10 *min, as indicated, for the Kézdy–Swinbourne plot*

For the first plot, Guggenheim recommended that τ should be several times $t_{\frac{1}{2}}$, the half-time of the reaction, i.e. the time required for half-completion. In contrast, Swinbourne suggested that values of τ in the range $0.5t_{\frac{1}{2}}$ to $t_{\frac{1}{2}}$ were the most suitable in the second type of plot. He also recommended

8

plotting the same data several times with different values of τ, in order to check that consistent values are obtained. This can also be done with the Guggenheim plot, of course, but is then much more laborious.

Both of these plots are insensitive to deviations from first-order kinetics, i.e. they can give apparently good straight lines in cases when first-order kinetics are not accurately obeyed. For this reason, neither plot should be used to test the order of a reaction, which should be established independently.

1.6 Influence of temperature on rate constants

From the earliest studies of reaction velocities, it has been evident that they are profoundly influenced by temperature. The most elementary consequence of this effect is that the temperature must always be controlled if meaningful results are to be obtained from kinetic experiments. However, with care, one can use temperature much more positively and by carrying out measurements at several temperatures, one can deduce important information about reaction mechanisms.

The studies of van't Hoff (1884) and Arrhenius (1889) form the starting point for all modern theories of the temperature dependence of rate constants. Harcourt (1867) had noted that the rates of many reactions approximately doubled for each 10°C increase in temperature, but van't Hoff and Arrhenius attempted to find a more exact relationship by comparing kinetic observations with the known properties of equilibrium constants. These constants, being thermodynamic quantities, were (and are) understood much more precisely than kinetic constants. Any equilibrium constant, K, varies with the absolute temperature, T, in accordance with the van't Hoff equation.

$$\frac{d \ln K}{dT} = \frac{\Delta H^\circ}{RT^2} \qquad (1.15)$$

where R is the gas constant and ΔH° is the standard enthalpy change in the reaction. By analogy with this equation, Arrhenius proposed a similar equation to describe the variation of a rate constant, k, with temperature:

$$\frac{d \ln k}{dT} = \frac{E_a}{RT^2} \qquad (1.16)$$

where E_a is the *activation energy*. Although this equation does not follow rigorously from equation 1.15, Arrhenius and later workers found it to be obeyed in practice by many reactions. Integration with respect to T gives

$$\ln k = \ln A - E_a/RT \qquad (1.17)$$

where $\ln A$ is a constant of integration. This form of the Arrhenius equation is the most convenient for the graphical treatment of results, as it shows that a plot of $\ln k$ against $1/T$ is a straight line, of slope $-E_a/R$. In practice, it is usually more convenient to plot $\log k$ against $1/T$, in which case the slope is $-E_a/2.303R$. This plot, which is illustrated in *Figure 1.2*, is known as an *Arrhenius plot*, and provides a simple method of evaluating E_a.

9

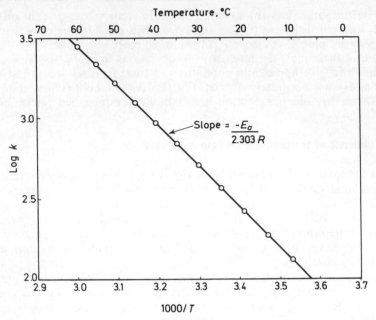

Figure 1.2 Arrhenius plot

In order to assess the meaning of the activation energy, equation 1.17 must be written as

$$k = A \exp(-E_a/RT)$$

The exponential term $\exp(-E_a/RT)$ is often called a Boltzmann term, because, according to Boltzmann's theory of the distribution of energies amongst molecules, the number of molecules in a mixture that have an energy in excess of E_a is proportional to $\exp(-E_a/RT)$. We can therefore interpret the Arrhenius equation to mean that molecules can take part in a reaction only if their energy exceeds some threshold value, the activation energy. In this interpretation, the constant A ought to be equal to the frequency of collisions of molecules, Z, at least for bimolecular reactions. For some simple reactions in the gas phase, such as the decomposition of hydrogen iodide, A is indeed equal to Z, but in general it is necessary to introduce a factor P:

$$k = PZ \exp(-E_a/RT)$$

and to assume that, in addition to colliding with sufficient energy, molecules must also be correctly oriented if they are to react. The factor P is then taken to be a measure of the probability that the correct orientation will be spontaneously adopted. This equation is now reasonably in accordance with modern theories of reaction rates, but for most purposes it is profitable to approach the same result from a different point of view, known as the transition-state theory, which is discussed in the next section.

10

1.7 Transition-state theory

The transition-state theory is derived largely from the work of Eyring (1935), and is so called because it attempts to relate the rates of chemical reactions to the thermodynamic properties of a particular high-energy state of the reacting molecules, known as the *transition state*, or *activated complex*. As a reacting system proceeds along a notional 'reaction co-ordinate', it must pass through a continuum of energy states, as illustrated in *Figure 1.3*, and at some stage it

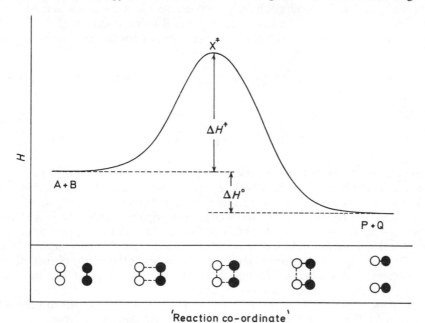

Figure 1.3 'Reaction profile' according to the transition-state theory: The diagrams along the abscissa indicate the meaning of the 'reaction co-ordinate' for a simple bimolecular reaction

must surpass a state of maximum energy. This maximum energy state is the transition state, and should be clearly distinguished from an intermediate, which is a metastable state of minimum energy (not *the* minimum in the popular sense, but *a* minimum in the mathematical sense). The reaction can be represented as follows:

$$A + B \underset{}{\overset{K^{\ddagger}}{\rightleftharpoons}} X^{\ddagger} \rightarrow P + Q$$

where $X^{\ddagger}$ is the transition state. Its concentration is assumed to be governed by the laws of thermodynamics, so that $[X^{\ddagger}] = K^{\ddagger}[A][B]$, where $K^{\ddagger}$ is given by

$$\Delta G^{\ddagger} = -RT \ln K^{\ddagger} = \Delta H^{\ddagger} - T \Delta S^{\ddagger}$$

where $\Delta G^{\ddagger}$, $\Delta H^{\ddagger}$ and $\Delta S^{\ddagger}$ are the free energy, enthalpy and entropy of forma-

11

tion, respectively, of the transition state from the reactants. The concentration of $X^{\ddagger}$ is therefore given by

$$[X^{\ddagger}] = [A][B]\exp(\Delta S^{\ddagger}/R)\exp(-\Delta H^{\ddagger}/RT)$$

As written, this equation, like any true thermodynamic equation, contains no information about time. In order to introduce time, we require quantum mechanical principles that are beyond the scope of this book (*see*, for example, Laidler, 1965), and the rate constant for the breakdown of $X^{\ddagger}$ can be shown to be RT/Nh, where R is the gas constant, N is Avogadro's number and h is Planck's constant. (The numerical value of RT/Nh is about $6.25 \times 10^{12}\ s^{-1}$ at 300 K.) Therefore, the second-order rate constant for the complete reaction is

$$k = \frac{RT}{Nh}\exp(\Delta S^{\ddagger}/R)\exp(-\Delta H^{\ddagger}/RT) \qquad (1.18)$$

Taking logarithms, and differentiating, we obtain

$$\frac{d \ln k}{dT} = (\Delta H^{\ddagger} + RT)/RT^2$$

Comparing this equation with the Arrhenius equation (equation 1.16), it can be seen that the activation energy, E_a, is not equal to $\Delta H^{\ddagger}$, but $\Delta H^{\ddagger} + RT$. Moreover, E_a is not strictly independent of temperature, so that the Arrhenius plot ought to be curved. The answer to this apparent anomaly is that the expected curvature is so slight that one would not normally expect to detect it. In fact, the variation in k that results from the factor T in equation 1.18 is trivial in comparison with variation in the exponential term. Nonetheless, the difference RT between E_a and $\Delta H^{\ddagger}$ amounts to about 2.6 kJ mol^{-1} at ordinary temperatures, and is not negligible.

As both A and E_a in equation 1.17 can readily be determined in practice from an Arrhenius plot, both $\Delta H^{\ddagger}$ and $\Delta S^{\ddagger}$ can be calculated, from

$$\Delta H^{\ddagger} = E_a - RT \approx E_a - 2490\ \text{J mol}^{-1}$$
$$\Delta S^{\ddagger} = R \ln(ANh/RT) - R \approx 19.1 \log A + 253\ \text{J mol}^{-1}\ \text{K}^{-1}$$

where the numerical equivalents are calculated by assuming that $T = 300$ K and that A is measured in s^{-1}. For values of T in the range 273–343 K (i.e. 0–70°C, the greatest range of temperature likely in an enzyme kinetic experiment), the variation in $\ln T$ is generally negligible in comparison with the experimental error in E_a and A. One can, in fact, avoid this slight source of error completely by plotting $\log(k/T)$ against $1/T$ instead of the usual Arrhenius plot of $\log k$ against $1/T$. In this case, it follows simply from equation 1.18 that the slope is $-\Delta H^{\ddagger}/2.303R$ and that the intercept is $\log(R/Nh) + \Delta S^{\ddagger}/2.303R$.

The enthalpy and entropy of activation of a chemical reaction provide valuable information about the nature of the transition state, and hence about the reaction mechanism. A large enthalpy of activation indicates that a large amount of stretching, squeezing or even breaking of chemical bonds is necessary for the formation of the transition state. For many chemical reactions that occur readily at room temperature (including many enzyme-

12

catalysed reactions), $\Delta H^{\ddagger}$ is found to be of the order of 50 kJ mol^{-1}. This value suggests that the energetic nature of the activation process is very similar for a wide variety of reactions, and it also accounts for Harcourt's observation that an increase in temperature of 10 °C often results in the rate being doubled.

The entropy of activation gives a measure of the inherent probability of the transition state, apart from energetic considerations. If $\Delta S^{\ddagger}$ is large and negative, the formation of the transition state requires that the reacting molecules adopt precise conformations and approach one another at a precise angle. As molecules vary widely in their conformational stability, i.e. their rigidity, and in their complexity, one might expect that values of $\Delta S^{\ddagger}$ would vary widely between different reactions. This does, in fact, occur. The molecules that are important in metabolic processes are often large and flexible, and so uncatalysed reactions between them are inherently unlikely, i.e. $-\Delta S^{\ddagger}$ is usually large.

Equation 1.18 shows that a catalyst can increase the rate of a reaction either by reducing $-\Delta S^{\ddagger}$ or by reducing $\Delta H^{\ddagger}$, or both. It is likely that both effects are important in enzymic catalysis, although in most cases it is not possible to obtain definite evidence of this because the uncatalysed reactions are too slow for $-\Delta S^{\ddagger}$ and $\Delta H^{\ddagger}$ to be measured.

2
Introduction to Enzyme Kinetics

2.1 Early studies

The rates of enzyme-catalysed reactions were first studied in the latter part of the nineteenth century by numerous workers. At that time, no enzyme was available in a pure form, methods of assay were primitive and the use of buffers to control pH had not been introduced. Moreover, it was customary to follow the course of the reaction over a period of time, in contrast to the more usual modern practice of measuring initial rates at various different initial substrate concentrations, which gives results that are generally easier to interpret. It is remarkable, therefore, how much progress was made.

Most of the early studies were concerned with enzymes from fermentation, particularly invertase, which catalyses the hydrolysis of sucrose:

$$\text{sucrose} + \text{water} \rightarrow \text{glucose} + \text{fructose}$$

O'Sullivan and Tompson (1890) studied this reaction, and made a number of important discoveries: they found that the reaction was highly dependent on the acidity of the mixture and that, provided that 'the acidity is in the most favourable proportion,' the rate was proportional to the amount of enzyme. The rate decreased as the substrate was consumed, and seemed to be proportional to the sucrose concentration, although there were slight deviations from the theoretical curve. At low temperatures, the enzyme showed an approximate doubling of rate for an increase in temperature of 10 °C. However, unlike most ordinary chemical reactions, the invertase-catalysed reaction displayed an apparent optimum temperature, above which the rate fell rapidly to zero. Invertase proved to be a true catalyst, as it was not destroyed or altered in the reaction, and a sample was still active after catalysing the hydrolysis of 100 000 times its weight of sucrose. Finally, the thermal stability of the enzyme was very much greater in the presence of its substrate, sucrose, than in its absence: 'invertase when in the presence of cane sugar will stand without injury a temperature fully 25 °C higher than in its absence. This is a very striking fact, and, as far as we can see, there is only

14

one explanation of it, namely, the invertase enters into combination with the sugar.' A similar conclusion had been reached by Wurtz (1880), who, in studying the papain-catalysed hydrolysis of fibrin, had observed a precipitate that he suggested might be a papain–fibrin compound that acted as an intermediate in the hydrolysis.

The idea of an *enzyme–substrate complex* was placed in a purely kinetic context by Brown (1892). In common with a number of other workers, he found that the rates of enzyme-catalysed reactions showed deviations from second-order kinetics. Initially, he showed that the rate of hydrolysis of sucrose in fermentation by live yeast appeared to be independent of sucrose concentration. The conflict between Brown's results with live yeast and those of O'Sullivan and Tompson with isolated invertase was not regarded as serious, because catalysis by isolated enzymes was regarded as fundamentally different from fermentation by living organisms. But Buchner's discovery (1897) that a cell-free (i.e. non-living) extract of yeast could catalyse alcoholic fermentation prompted Brown (1902) to re-examine his earlier results. After confirming that they were correct, he showed that similar results could be obtained with purified invertase and suggested that the enzyme–substrate complex mechanism placed a limit on the rate that could be achieved. Provided that the complex existed for a brief instant of time before breaking down to products, then a maximum rate would be reached when the substrate concentration was sufficiently high to convert all of the enzyme into complex, according to the law of mass action. At lower concentrations of substrate, the rate at which complex was formed would become significant, and so the rate of hydrolysis would be dependent on substrate concentration.

Henri (1902, 1903) criticized Brown's model on the grounds that it assumed a fixed lifetime for the enzyme–substrate complex between its abrupt creation and decay. He proposed instead a mechanism that was conceptually very similar to Brown's but that was expressed in more precise mathematical and chemical terms, with an equilibrium between the free enzyme and the enzyme–substrate and enzyme–product complexes. Henri showed that it did not matter whether one assumed either or both of these complexes to be intermediates in the reaction: identical kinetics would be observed whether products were formed by the breakdown of the enzyme–substrate complex:

$$E + S \rightleftharpoons ES \rightarrow E + P$$

or whether the free enzyme were the active species in a second-order reaction:

$$\begin{array}{c} ES \\ \uparrow\downarrow \\ E + S \rightarrow E + P \end{array}$$

This latter possibility is not now regarded as very plausible, and has only rarely been considered (e.g. by Viale, 1970), but it was in accord with the ideas of the time to assume that a catalyst acted merely by its presence. Moreover, the principle that two or more mechanisms may require the same form of rate equation (sometimes called *homeomorphism*) is important, and should always be borne in mind when kinetic experiments are interpreted.

It is of interest that Henri allowed for an enzyme–product complex in his

15

formulation, and that the concept of competitive inhibition by product is implicit in the rate equation that he gave:

$$v = \frac{\dfrac{k}{K_s} e_0 s}{1 + \dfrac{s}{K_s} + \dfrac{p}{K_p}} \tag{2.1}$$

in which v is the velocity, e_0 is the total enzyme concentration, s and p are the concentrations of free substrate and product, respectively, and k, K_s and K_p are constants. If p is set equal to zero, this equation defines the values of the initial rates for different substrate concentrations. The more general form used by Henri was more appropriate to his experiments because, in common with other investigators of that time, he followed the course of the reaction over an extended period. Nonetheless, his equation is incomplete, because it takes no account of the overall back reaction: it implies that the accumulation of product retards the forward reaction, but does not prevent it from going to completion. A more realistic equation (e.g. equation 2.18 below) makes allowance for the possibility that the reaction is incomplete at equilibrium.

2.2 Work of Michaelis and Menten

Although Brown and Henri reached essentially correct conclusions, they did so on the basis of experiments that were open to serious objections. O'Sullivan and Tompson experienced great difficulty in obtaining coherent results until they realized the importance of acid concentration. Brown prepared the enzyme in a different way and found the addition of acid to be unnecessary (presumably his solutions were weakly buffered by the natural components of the yeast), and Henri did not discuss the problem. Apart from O'Sullivan and Tompson, the early investigators of invertase made no allowance for the mutarotation of the glucose produced in the reaction, although this undoubtedly affected the results.

With the introduction of the concept of hydrogen ion concentration, expressed by the logarithmic scale of pH (Sørensen, 1909), Michaelis and Menten (1913) realized the necessity for carrying out definitive experiments with invertase. They controlled the pH of the reaction by the use of acetate buffers, they allowed for the mutarotation of the product and they measured *initial rates* of the reaction at different substrate concentrations. If initial rates are used, the back reaction and other effects of product can legitimately be ignored, so that a much simpler rate equation can be used. In spite of these refinements, Michaelis and Menten obtained results in good agreement with those of Henri, and they acknowledged their debt to him and to Brown in the mechanism that they suggested:

$$E + S \rightleftharpoons ES \rightarrow E + P$$

Like Henri, they assumed that the reversible first step was sufficiently rapid for it to be represented by an equilibrium constant, $K_s = es/x$, where x is the concentration of the intermediate, ES, so that $x = es/K_s$. The instantaneous concentrations of free enzyme and substrate are not directly measurable, however, and so they must be expressed in terms of the initial, measured, concentrations, e_0 and s_0, using the relationships

$$e_0 = e + x$$
$$s_0 = s + x$$

From the first of these, x cannot be greater than e_0, and so, provided that s_0 is very large compared with e_0, it must also be very large compared with x. Thus $s = s_0$ with good accuracy. Then the expression for x becomes

$$x = (e_0 - x)s/K_s$$

which can be rearranged to give

$$x = \frac{e_0}{(K_s/s) + 1}$$

The second step in the reaction, $ES \rightarrow E + P$, is a simple first-order reaction, with a rate constant k_{+2}, so that

$$v = k_{+2}x = \frac{k_{+2}e_0}{(K_s/s) + 1} = \frac{k_{+2}e_0 s}{K_s + s} \qquad (2.2)$$

This equation is identical with the equation given by Henri (equation 2.1) in the special case when $p = 0$.

Michaelis and Menten showed that this theory, and equation 2.2, could account accurately for their results with invertase. Because of the definitive nature of their experiments, which have served as a standard for almost all subsequent enzyme kinetic measurements, Michaelis and Menten are regarded as the founders of modern enzymology, and equation 2.2 (in its modern form, equation 2.8, below) is generally known as the *Michaelis–Menten equation*.

At about the same time, Van Slyke and Cullen (1914) obtained similar results with the enzyme urease. They assumed a similar mechanism, with the important difference that the first step was assumed to be irreversible.

$$E + S \xrightarrow{k_{+1}} ES \xrightarrow{k_{+2}} E + P$$
$$e_0 - x \qquad s \qquad x \qquad p$$

In this case, of course, x cannot be represented by an equilibrium constant; instead, we have

$$dx/dt = k_{+1}(e_0 - x)s - k_{+2}x$$

Van Slyke and Cullen implicitly assumed that the intermediate concentration was constant, i.e. $dx/dt = 0$, and so

$$x = \frac{k_{+1}e_0 s}{k_{+2} + k_{+1}s}$$

Therefore,

$$v = k_{+2}x = \frac{k_{+1}k_{+2}e_0 s}{k_{+2}+k_{+1}s} = \frac{k_{+2}e_0 s}{(k_{+2}/k_{+1})+s} \tag{2.3}$$

This equation is of the same form as equation 2.2, and empirically indistinguishable from it.

At about the same time as these developments were taking place in the understanding of enzyme catalysis, similar conclusions were being reached in the study of the adsorption of gases by solids (Langmuir, 1916, 1918). Langmuir's treatment was considerably more general, but the case that he referred to as *simple adsorption* corresponds closely to the type of binding assumed by Henri and by Michaelis and Menten. He recognized the similarity between solid surfaces and enzymes, although he imagined the whole surface of an enzyme to be 'active,' rather than limited areas or *active sites*. Hitchcock (1926) pointed out the similarity between the equations for the binding of ligands to solid surfaces and to proteins, and the logical process was completed by Lineweaver and Burk (1934), who extended Hitchcock's ideas to catalysis.

2.3 Steady-state treatment

The formulation of Michaelis and Menten, which treats the first step of enzyme catalysis as an equilibrium, and that of Van Slyke and Cullen, which treats it as irreversible, both make unwarranted and unnecessary assumptions about the rate constants. As we have seen, both formulations lead to the same form of the rate equation, and Briggs and Haldane (1925) examined a generalized mechanism that includes both special cases:

$$E + S \underset{k_{-1}}{\overset{k_{+1}}{\rightleftharpoons}} ES \overset{k_{+2}}{\longrightarrow} E + P$$

$$e_0 - x \qquad s \qquad x$$

In this case,

$$dx/dt = k_{+1}(e_0 - x)s - k_{-1}x - k_{+2}x \tag{2.4}$$

If it is assumed that a steady state is achieved in which the concentration of the intermediate is constant, i.e. $dx/dt = 0$, then

$$k_{+1}e_0 s = (k_{+1}s + k_{-1} + k_{+2})x \tag{2.5}$$

Therefore,

$$x = \frac{k_{+1}e_0 s}{k_{+1}s + k_{-1} + k_{+2}} \tag{2.6}$$

and so

$$v = k_{+2}x = \frac{k_{+1}k_{+2}e_0 s}{k_{+1}s + k_{-1} + k_{+2}} = \frac{k_{+2}e_0 s}{\dfrac{k_{-1} + k_{+2}}{k_{+1}} + s} \tag{2.7}$$

18

This equation can be written in the more general form:

$$v = Vs/(K_m + s) \tag{2.8}$$

where K_m, known as the *Michaelis constant*, is defined as $(k_{-1} + k_{+2})/k_{+1}$, and V, known as the *maximum velocity*, is defined as $k_{+2}e_0$. Equation 2.8 is the fundamental equation of enzyme kinetics, and is usually called the *Michaelis–Menten equation*. It applies to many mechanisms more complex than the Michaelis–Menten mechanism, but with more complex definitions of K_m and V. In practice, therefore, it cannot be assumed that K_m can be expressed simply as $(k_{-1} + k_{+2})/k_{+1}$, or V as $k_{+2}e_0$. V is not a fundamental property of an enzyme, as it depends upon enzyme concentration. Provided that the enzyme concentration is known, it is advantageous to define a quantity k_{cat}, the '*catalytic constant*' or '*turnover number*,' as V/e_0. For the Michaelis–Menten mechanism, k_{cat} is identical with k_{+2}, but in general the more non-committal notation k_{cat} is preferable.

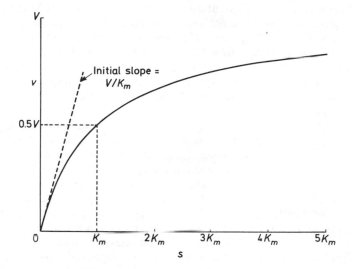

Figure 2.1 *Plot of initial velocity,* v, *against substrate concentration,* s, *for a reaction obeying the Michaelis–Menten equation: This plot was* not *used or advocated by Michaelis and Menten*

The graphical form of equation 2.8 is shown in *Figure 2.1*. The curve is a rectangular hyperbola, with asymptotes $s = -K_m$ and $v = V$. At very small values of s, v is directly proportional to s:

$$v \approx Vs/K_m$$

so that the reaction is apparently first order in s. It is instructive to realize that V/K_m has this fundamental meaning as the rate constant for the reaction $E + S \rightarrow E + P$ at low substrate concentrations, and that it should not be regarded solely as the ratio of V to K_m. When s is equal to K_m, the velocity is 'half-maximal,' i.e. $v = 0.5V$. At very high values of s, v approximates to V, and the reaction is apparently zero order in s; under these conditions, the enzyme is said to be saturated. Actually, v approaches V rather slowly, and even when $s = 10K_m$, v is still only $0.91V$. For this reason, V often cannot be found by direct measurement, but must be estimated from the velocities observed at sub-saturating values of s. It is almost impossible to estimate the asymptotes of a rectangular hyperbola accurately, because of a tendency to draw them too close to the curve, unless the results are plotted in a different way. Methods for doing this are discussed in Section 2.5.

It is tempting to assume that K_m can be taken as a measure of the true binding constant, K_s, in practice, i.e. to assume that k_{+2} is negligible in comparison with k_{-1}. This is a very dubious assumption unless supported by other evidence, and should be made with great caution. There are very few enzymes for which the individual values of k_{-1} and k_{+2} are known (for a review, see Eigen and Hammes, 1963) and indeed, horseradish peroxidase, the first enzyme for which the individual rate constants were measured, was found to have $k_{+2} \gg k_{-1}$ (Chance, 1943), so that it agreed much better with the assumptions of Van Slyke and Cullen than with those of Michaelis and Menten. More important, there are many mechanisms, more complicated than that assumed by Briggs and Haldane, that generate a steady-state rate equation of the form of equation 2.8. In such cases, more complex expressions for K_m are required, which do not necessarily simplify to K_s under plausible conditions. In fact, K_m cannot even be taken as an upper limit for K_s (Dalziel, 1962), as it could legitimately be if the Briggs–Haldane treatment were completely general.

In view of this discussion, the reader may well wonder if there is any point in measuring K_m, if it cannot be used as a measure of binding. There are, in fact, several reasons for measuring K_m. Firstly, in order to understand clearly the nature of a complex mechanism, it is usually necessary to express complex effects in simple terms, which is most easily achieved by determining how the basic kinetic parameters K_m, V and V/K_m vary with the experimental conditions. Secondly, K_m is useful purely as a predictive parameter, permitting the design of valid enzyme assays. In the design of an assay, it is desirable that the measured rate should depend only on the enzyme concentration and not on small errors in the substrate concentrations. Although ideally each substrate should therefore be saturating (i.e. at infinite concentration), in practice a concentration of $10K_m$ or greater is sufficient to make the rate insensitive to errors in the substrate concentration. Finally, if it is taken in conjunction with measurements of inhibitor binding constants (which are often true thermodynamic quantities; see Sections 3.7 and 4.2) for analogues of the substrate, K_m *can* sometimes be interpreted cautiously as a measure of K_s. For example, all three stereoisomers of the pepsin substrate acetyl-L-phenylalanyl-L-phenylalanine have competitive inhibition constants (K_i) similar in magnitude to K_m for this substrate (Knowles, Sharp and Greenwell, 1969), and in this instance it would be perverse not to regard K_m as an indication of K_s. Of course, this does not prove that $k_{+2} \ll k_{-1}$, but it does make it rather unlikely that $k_{+2} \gg k_{-1}$. K_m is also likely to be equal to K_s if it remains constant under variations in pH or effector concentration that cause variations in V, as is discussed in Section 6.4.

2.4 Validity of the steady-state assumption

The steady-state assumption was introduced by Bodenstein (1913) in order to account for the rates of certain photochemical reactions. As we have seen, it was used implicitly by Van Slyke and Cullen (1914) and explicitly by Briggs and Haldane (1925) to derive rate equations for enzyme-catalysed reactions. None of these workers justified it rigorously, and unfortunately a tendency has arisen to regard it as being axiomatic or at least obviously true. In fact, it is *not* accurately true for some uncatalysed reactions such as

$$A + B \rightleftharpoons C \rightarrow D$$

where it appears to be just as plausible at first sight as in the Briggs–Haldane case. As the steady-state assumption has become the central assumption underlying the derivation of most of the rate equations used in enzyme chemistry, it is important to realize that it can be derived fairly simply, at least in the case of the Michaelis–Menten mechanism.

If the steady-state assumption is not made, then the differential equation (equation 2.4) cannot be solved by putting $dx/dt = 0$, but instead must be integrated as follows:

$$\int \frac{dx}{k_{+1}e_0 s - (k_{+1}s + k_{-1} + k_{+2})x} = \int dt$$

Therefore,

$$\frac{\ln[k_{+1}e_0 s - (k_{+1}s + k_{-1} + k_{+2})x]}{-(k_{+1}s + k_{-1} + k_{+2})} = t + \alpha$$

At the instant when the reaction starts, there can be no intermediate, i.e. $x = 0$ when $t = 0$, and so

$$\alpha = \frac{\ln[k_{+1}e_0 s]}{-(k_{+1}s + k_{-1} + k_{+2})}$$

giving

$$\ln\left[\frac{k_{+1}e_0 s - (k_{+1}s + k_{-1} + k_{+2})x}{k_{+1}e_0 s}\right] = -(k_{+1}s + k_{-1} + k_{+2})t$$

Therefore,

$$1 - \frac{(k_{+1}s + k_{-1} + k_{+2})x}{k_{+1}e_0 s} = \exp[-(k_{+1}s + k_{-1} + k_{+2})t]$$

Hence,

$$x = \frac{k_{+1}e_0 s\{1 - \exp[-(k_{+1}s + k_{-1} + k_{+2})t]\}}{k_{+1}s + k_{-1} + k_{+2}}$$

and

$$v = \frac{Vs\{1 - \exp[-(k_{+1}s + k_{-1} + k_{+2})t]\}}{K_m + s} \tag{2.9}$$

where V and K_m are defined as before, i.e. as $k_{+2}e_0$ and $(k_{-1} + k_{+2})/k_{+1}$, respectively. Inspection of equation 2.9 shows that it reduces to the steady-state expression (equation 2.8) when t is large, as the exponential term then disappears. How large t must be for this to happen will depend on the magnitude of $(k_{+1}s + k_{-1} + k_{+2})$, but the fact that most enzyme-catalysed reactions are found to obey equation 2.8, except during the first fraction of a second after mixing, can be taken as evidence that this quantity is much greater than $1\ s^{-1}$ in most instances.

An equation of the form of equation 2.9 was derived by Laidler (1955) as a special case of a much more general treatment than is given here. He considered the effect of allowing the reaction to proceed for long enough for the decrease in substrate concentration to be significant, so that it would not be valid to regard s as a constant equal to s_0, the initial value of s. He found that a steady state was achieved in which

$$x = \frac{k_{+1}e_0(s_0 - p)}{k_{-1} + k_{+2} + k_{+1}(s_0 - p)} \tag{2.10}$$

and so

$$v = \frac{k_{+1}k_{+2}e_0(s_0 - p)}{k_{-1} + k_{+2} + k_{+1}(s_0 - p)} = \frac{V(s_0 - p)}{K_m + s_0 - p} \tag{2.11}$$

Equations 2.10 and 2.11 are the same as equations 2.6 and 2.8 except for the replacement of s with $s_0 - p$. It may seem contradictory to refer to a steady state in which x is represented by an expression in p, which varies with time. This paradox is resolved by consideration of the fact that the time dependence of x as expressed by equation 2.10 is far less than that in the transient phase, i.e. the period before the steady state is established. The argument used by Briggs and Haldane is very little affected by replacing the assumption that

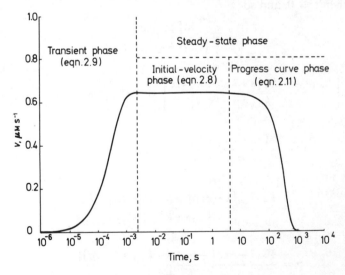

Figure 2.2 Time course of a reaction proceeding by the Michaelis–Menten mechanism: $k_{+1} = 10^7\ \text{M}^{-1}\,\text{s}^{-1}$; $k_{-1} = 1000\ s^{-1}$; $k_{+2} = 100\ s^{-1}$; $k_{-2} = 0$; $s_0 = 2 \times 10^{-4}\ \text{M}$; $e_0 = 10^{-8}\ \text{M}$

$dx/dt = 0$ with an assumption that dx/dt is very small: equation 2.5 becomes merely a good approximation instead of an exact statement.

We can summarize these results by saying that an enzyme-catalysed reaction normally proceeds through three definite phases, well separated on the time scale, as illustrated in *Figure 2.2*. The first, the transient phase, requires equations similar to equation 2.9 for its description, and is discussed in more detail in Chapter 9. The second, the initial-velocity phase, is the one in which the velocity is virtually constant, and has been by far the most studied of the three phases since the work of Michaelis and Menten; the bulk of this book is devoted to it. The final phase, in which the substrate and product concentrations change significantly and the velocity decreases to zero, requires equations similar to equation 2.11 (usually in integrated form), and is discussed in Chapter 8.

We have hitherto assumed that $s_0 \gg e_0$ (the most common situation in practice for studies of the steady state) in deriving rate equations. Laidler (1955) examined the effect of the steady-state assumption under less restricted conditions. He found that a slightly modified form of equation 2.11, as follows:

$$v = \frac{V(s_0 - p)}{K_m + e_0 + s_0 - p}$$

was valid provided that any one (or more) of the following conditions was satisfied:

(1) $s_0 \gg e_0$; (2) $e_0 \gg s_0$; (3) $k_{-1} + k_{+2} \gg k_{+1}e_0$; or (4) $k_{-1} + k_{+2} \gg k_{+1}s_0$.

As the experimenter normally has wide freedom in the choice of values of e_0 and s_0, it is usually possible to ensure that at least one of these conditions is satisfied.

Finally, it should be noted that this discussion does not provide for the effect of product inhibition, which may occur. This is considered in Section 2.7.

2.5 Graphical representation of the Michaelis–Menten equation

If a series of initial velocities at different substrate concentrations is measured, it is desirable to present the results graphically, so that the values of the kinetic parameters and the precision of the experiment can be estimated. The most obvious way of plotting equation 2.8 is to plot v against s, generating a rectangular hyperbola, with $v = V$ as the asymptote for v and $s = -K_m$ as the asymptote for s, as was shown in *Figure 2.1*. This is a most unsatisfactory plot in practice, because it is difficult to draw rectangular hyperbolas accurately; it is difficult to estimate asymptotes accurately (there is a strong tendency to draw asymptotes too close to the curve); it is difficult to perceive the relationship between a family of hyperbolas; and it is difficult to detect deviations from the expected curve if they occur. These disadvantages were recognized by Michaelis and Menten (1913), who instead plotted v against

23

log s. This plot gives a symmetrical S-shaped curve, as shown in *Figure 2.3*,

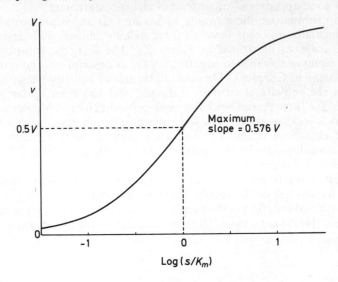

Figure 2.3 Determination of V *and* K$_m$ *by the method of Michaelis and Menten: The plot is almost straight over an appreciable range of velocities, and so the maximum slope can be estimated fairly easily*

which has maximum slope when $s = K_m$. Differentiation of equation 2.8 shows that

$$\frac{dv}{d \ln s} = \frac{K_m V s}{(K_m + s)^2}$$

or

$$\frac{dv}{d \log s} = \frac{2.303 K_m V s}{(K_m + s)^2}$$

The maximum slope occurs when $s = K_m$, and is equal to $2.303V/4$, i.e. $0.576V$. Hence, V can be estimated by dividing the maximum slope by 0.576. Michaelis and Menten then estimated K_m as the value of s where the velocity was half of V. This can be achieved more accurately than estimating the point at which the maximum slope occurs. This plot is not usually used today, but it is of interest for several reasons: it emphasizes the relationship between the saturation of a protein and the ionization of an acid, which is usually represented by a titration curve in which the degree of ionization is plotted against pH; it is used in a more general form to represent the saturation of proteins that have several binding sites; it does not suffer from statistical bias in the way that the linear plots (which will be considered next) do, although it does tend to weight the observations in the middle of the range more than those at the extremes; and, finally, it is of historical interest.

Most workers since Lineweaver and Burk (1934) have preferred to re-write the Michaelis–Menten equation in a form that permits the results to be plotted as a straight line. This can be achieved in three ways:

$$\frac{s}{v} = \frac{K_m}{V} + \frac{s}{V} \tag{2.12}$$

$$v = V - \frac{K_m v}{s} \tag{2.13}$$

$$\frac{1}{v} = \frac{1}{V} + \frac{K_m}{Vs} \tag{2.14}$$

In the first case, a plot of s/v against s gives a straight line of slope $1/V$ and intercepts K_m/V on the s/v axis and $-K_m$ on the s axis. Similarly, straight-line plots are obtained from equations 2.13 and 2.14 by plotting v against v/s and $1/v$ against $1/s$, respectively. These three plots are illustrated in *Figures 2.4–2.6*.

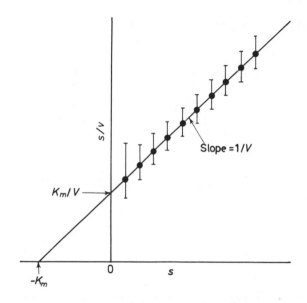

Figure 2.4 Plot of s/v against s, with error bars of ±0.05V in v

An equation of the form of equation 2.12 was given by Langmuir (1918) in connection with the adsorption of gases on to solid surfaces, and was first given in the context of enzyme kinetics by Hanes (1932), who used it in the analysis of his results, although he did not present them graphically. The other two linear transformations, and all three plots, were introduced by Woolf (1932), but became widely known and used as a result of the work of Lineweaver and Burk (1934), Eadie (1942) and Hofstee (1952). By far the most widely used has been the *double-reciprocal plot* of $1/v$ against $1/s$, commonly referred to as the Lineweaver–Burk plot, but because it is also by far the worst plot, its use should be discouraged.

25

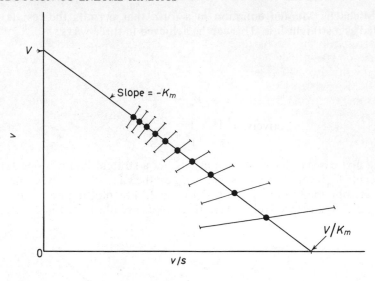

Figure 2.5 Plot of v *against* v/s, *with error bars of* ± 0.05V *in* v *(Eadie–Hofstee plot)*

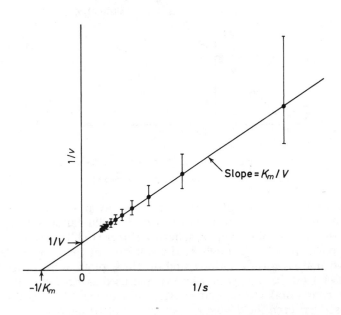

Figure 2.6 Plot of 1/v *against* 1/s, *with error bars of* ± 0.05V *in* v *(Lineweaver–Burk or double-reciprocal plot)*

26

Until 1961, most of the discussion of the relative merits of the three linear plots was concerned with their convenience and with their ability to show departures from the Michaelis–Menten equation clearly. So long as the discussion is confined to these aspects, there is little to choose between them, and the situation was cogently summed up by Dixon and Webb (*see* Hofstee, 1959). More recently, the discussion has been concerned with the statistical bias inherent in all linear transformations of the Michaelis–Menten equation, largely as a result of the studies of Wilkinson (1961) and Johansen and Lumry (1961)*. Their results and the conclusions to be drawn from them are discussed in greater detail in Chapter 10, but a few important points are mentioned here. For definitive work, it is unwise to use any plot, linear or otherwise, for estimating the parameters. Instead, a computer program should be used, as described in Chapter 10. For illustrative purposes, one of the linear plots can be used. It is sometimes argued that as all three of these plots are unsatisfactory, one might as well use the most familiar, namely the double-reciprocal plot. However, this proposal is contrary to the principle that justice should not merely be done, but should be seen to be done: if an experiment is illustrated with a double-reciprocal plot in which the line shown is derived from an independent unbiassed calculation, the line will generally appear incorrect on account of large deviations of some of the points at low substrate concentrations. This problem also arises to some extent with the other two linear plots, but much less severely. The plot of v against v/s has the disadvantage that v, usually regarded as the dependent variable, appears in both co-ordinates. On balance, the plot of s/v against s is the most satisfactory of the three.

Eisenthal and Cornish-Bowden (1974) have recently described a completely different method of plotting enzyme kinetic results, which they call the *direct linear plot*. Instead of writing the Michaelis–Menten equation in the usual way to show the dependence of v on s, they rearrange it to show the dependence of V on K_m:

$$V = v + \frac{v}{s} K_m \tag{2.15}$$

Thus, for any values of s and v, it is possible to plot V against K_m, as a straight line with slope v/s, intercept $-s$ on the K_m axis, and intercept v on the V axis. This straight line relates all values of V and K_m that satisfy the particular values of s and v exactly. If straight lines are drawn in this way for several observations, they should intersect at a common point, the co-ordinates of which give the only values of V and K_m that satisfy all of the observations, as illustrated in *Figure 2.7*. In a real experiment, the point of intersection is less well defined than that shown in *Figure 2.7* on account of experimental error, but it is easy to find the best point, where the lines crowd closest together. It is also easy to recognize particularly bad observations, as they give rise to

---

* Actually, Hanes (1932) noted the dangers of statistical bias in the very first paper in which a linear transformation of the Michaelis–Menten equation was used, but he felt them to be of minor importance. As he used equation 2.12, which is not seriously biassed, his assessment was not far from the truth. Unfortunately, the same cannot be said of the very severely biassed equation 2.14 and the double-reciprocal plot derived from it.

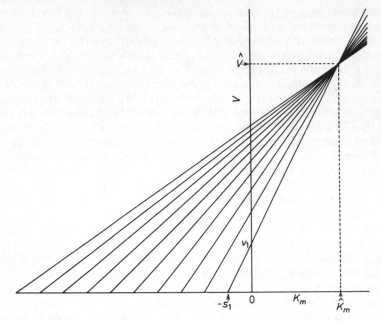

Figure 2.7 Direct linear plot of V *against* K_m: *Each line represents one observation, and is drawn with intercepts* $-s$ *on the abscissa and* v *on the ordinate. The point of intersection gives the co-ordinates of the best-fit values,* $\hat{K}_m$ *and* $\hat{V}$

lines that plainly disagree with the majority. The most obvious advantage of the direct linear plot is that it requires no calculation—not even reciprocals need to be determined. It also has certain valuable statistical properties, which are discussed in Section 10.9.

2.6 Reversible Michaelis–Menten mechanism

All reactions are reversible in principle, and many of the reactions of importance in biochemistry are also reversible in practice, in the sense that significant amounts of both substrates and products exist in the equilibrium mixture. It is evident, therefore, that the Michaelis–Menten mechanism, as given, is incomplete, and that allowance should be made for the reverse reaction:

$$E \; + \; A \underset{k_{-1}}{\overset{k_{+1}}{\rightleftharpoons}} EA \underset{k_{-2}}{\overset{k_{+2}}{\rightleftharpoons}} E \; + \; P \qquad (2.16)$$

$$e_0 - x \qquad a \qquad x \qquad e_0 - x \qquad p$$

(When discussing mechanisms in which we are interested in more than one substrate, it is convenient not to use the symbol S for any particular substrate,

28

but to reserve it for substrates in general. In these cases, A, B,... are used for substrates of the forward reaction and P, Q,... for substrates of the reverse reaction.) The steady-state assumption is now expressed by

$$dx/dt = k_{+1}(e_0-x)a+k_{-2}(e_0-x)p-(k_{-1}+k_{+2})x = 0$$

Rearranging, we obtain

$$x = \frac{k_{+1}e_0a+k_{-2}e_0p}{k_{-1}+k_{+2}+k_{+1}a+k_{-2}p}$$

The *net* rate of production of P is the difference between the forward and reverse reactions:

$$v = k_{+2}x-k_{-2}(e_0-x)p$$

Substitution for x in this expression gives

$$v = \frac{k_{+1}k_{+2}e_0a-k_{-1}k_{-2}e_0p}{k_{-1}+k_{+2}+k_{+1}a+k_{-2}p} \tag{2.17}$$

The special case $p = 0$ gives the same equation as obtained before, i.e. equation 2.7 (although a must be replaced with a_0, because the initial-rate condition is strictly satisfied only at zero time), while the special case $a = 0$ gives the corresponding expression for the initial rate of the reverse reaction:

$$v = \frac{-k_{-1}k_{-2}e_0p_0}{k_{-1}+k_{+2}+k_{-2}p_0}$$

As with the expression for the forward reaction, this equation is of the form of the Michaelis–Menten equation (equation 2.8) and we can define

$$V^r = k_{-1}e_0$$
$$K_m^P = (k_{-1}+k_{+2})/k_{-2}$$

for the reverse reaction, analogous to the definitions of V and K_m for the forward reaction:

$$V^f = k_{+2}e_0$$
$$K_m^A = (k_{-1}+k_{+2})/k_{+1}$$

Using these four definitions, equation 2.17 can be re-written as

$$v = \frac{\dfrac{V^fa}{K_m^A} - \dfrac{V^rp}{K_m^P}}{1 + \dfrac{a}{K_m^A} + \dfrac{p}{K_m^P}} \tag{2.18}$$

This equation can be regarded as the general reversible form of the Michaelis–Menten equation. It has the advantage over equation 2.17 that it does not imply a particular mechanism and can be regarded as purely empirical: there are many mechanisms more complicated than equation 2.16 that nonetheless generate equation 2.18. The most important of these is the more realistic

29

reversible mechanism in which the conversion of A into P and the release of P from the enzyme are treated as distinct reactions:

$$E + A \underset{k_{-1}}{\overset{k_{+1}}{\rightleftharpoons}} EA \underset{k_{-2}}{\overset{k_{+2}}{\rightleftharpoons}} EP \underset{k_{-3}}{\overset{k_{+3}}{\rightleftharpoons}} E + P \tag{2.19}$$

$$e_0 - x - y \quad a \qquad x \qquad y \quad e_0 - x - y \quad p$$

The steady-state rate equation can be derived by setting the rates of change of both intermediate concentrations to zero:

$$dx/dt = k_{+1}a(e_0 - x - y) + k_{-2}y - (k_{-1} + k_{+2})x = 0$$
$$dy/dt = k_{-3}p(e_0 - x - y) + k_{+2}x - (k_{-2} + k_{+3})y = 0$$

and solving the resulting simultaneous equations for x and y:

$$x = \frac{k_{+1}(k_{-2} + k_{+3})e_0 a + k_{-2}k_{-3}e_0 p}{k_{-1}k_{-2} + k_{-1}k_{+3} + k_{+2}k_{+3} + k_{+1}(k_{-2} + k_{+2} + k_{+3})a + (k_{-1} + k_{-2} + k_{+2})k_{-3}p}$$

$$y = \frac{k_{+1}k_{+2}e_0 a + (k_{-1} + k_{+2})k_{-3}e_0 p}{k_{-1}k_{-2} + k_{-1}k_{+3} + k_{+2}k_{+3} + k_{+1}(k_{-2} + k_{+2} + k_{+3})a + (k_{-1} + k_{-2} + k_{+2})k_{-3}p}$$

The net rate is the difference between the forward and reverse rates for any step:

$$v = k_{+2}x - k_{-2}y$$

$$= \frac{k_{+1}k_{+2}k_{+3}e_0 a - k_{-1}k_{-2}k_{-3}e_0 p}{k_{-1}k_{-2} + k_{-1}k_{+3} + k_{+2}k_{+3} + k_{+1}(k_{-2} + k_{+2} + k_{+3})a + (k_{-1} + k_{-2} + k_{+2})k_{-3}p}$$

This equation is of the same form as equation 2.18, but the definitions of the kinetic parameters are now

$$V^f = \frac{k_{+2}k_{+3}e_0}{k_{-2} + k_{+2} + k_{+3}}$$

$$V^r = \frac{k_{-1}k_{-2}e_0}{k_{-1} + k_{-2} + k_{+2}}$$

$$K_m^A = \frac{k_{-1}k_{-2} + k_{-1}k_{+3} + k_{+2}k_{+3}}{k_{+1}(k_{-2} + k_{+2} + k_{+3})} \tag{2.20}$$

$$K_m^P = \frac{k_{-1}k_{-2} + k_{-1}k_{+3} + k_{+2}k_{+3}}{(k_{-1} + k_{-2} + k_{+2})k_{-3}} \tag{2.21}$$

In spite of its complex appearance, equation 2.20 simplifies to $K_m^A = k_{-1}/k_{+1} = K_s^A$, the true dissociation constant of EA, in the event that k_{+2} is small in comparison with $(k_{-2} + k_{+3})$. Similarly, $K_m^P \approx K_s^P$ if k_{-2} is small in comparison with $(k_{-1} + k_{+2})$. Both of these approximations can be simultaneously true if $k_{+2} \ll k_{+3}$ and $k_{-2} \ll k_{-1}$, in other words if the interconversion of EA and EP is rate-determining in both directions.

When a reaction is at equilibrium, the net velocity must be zero and, consequently, if a_∞ and p_∞ are the equilibrium values of a and p, it follows from equation 2.18 that

$$\frac{V^f a_\infty}{K_m^A} - \frac{V^r p_\infty}{K_m^P} = 0$$

and so

$$\frac{V^f K_m^P}{V^r K_m^A} = \frac{p_\infty}{a_\infty} = K$$

where K is the equilibrium constant of the reaction. This is an important result, and is known as the *Haldane relationship* (Haldane, 1930). It is true for any mechanism that is described by equation 2.18, not merely for the simple Michaelis–Menten mechanism. More complex rate equations, such as those which involve several substrates, require more complex relationships, but in all instances at least one relationship of this type must exist between the kinetic parameters and the equilibrium constant.

The Haldane relationship is actually a generalization of an equation used earlier by von Euler and Josephson (1924) to compare the kinetic parameters for the enzymic hydrolysis of β-methylglucose with the equilibrium constant of the reaction. Its importance was largely unrecognized, however, until it was applied by Bock and Alberty (1953) to the kinetic parameters of fumarase. Since then, it has become a standard precaution to check that kinetic results are in accordance with the Haldane relationship, provided, of course, that it is possible to follow the reaction in both forward and reverse directions.

2.7 Product inhibition

Product inhibition is simply a special case of inhibition, which is discussed in detail in Chapter 4, but because it follows very naturally from the previous section it is convenient to discuss it briefly here. When equation 2.18 applies, the rate must decrease as product accumulates, even if the decrease in substrate concentration is negligible, because the negative term in the numerator becomes relatively more important as equilibrium is approached, and because the third term in the denominator increases. In any reaction, the negative term in the numerator can be significant only if the reaction is significantly reversible. Now, in many essentially irreversible reactions, such as the classic example of the invertase-catalysed hydrolysis of sucrose, product inhibition is significant. This is compatible with the simplest mechanism (equation 2.16) only if the *first* step is irreversible and the second is not. This does not seem very likely, at least as a general phenomenon. On the other hand, the two-intermediate mechanism (equation 2.19) predicts that product inhibition can occur in an irreversible reaction if it is the second step that is irreversible. In such a case, the accumulation of product causes the enzyme to be sequestered as the EP complex. For an irreversible reaction, equation 2.18 then becomes

$$v = \frac{V^f a/K_m^A}{1 + \dfrac{a}{K_m^A} + \dfrac{p}{K_s^P}} = \frac{V^f a}{K_m^A(1 + p/K_s^P) + a} \qquad (2.22)$$

K_m^P can legitimately be written as K_s^P if the reaction is irreversible, because if k_{-2} approximates to zero it is necessarily small compared with $(k_{-1} + k_{+2})$ (cf. equation 2.21).

Of course, the effect of added product should be the same as the effect of accumulated product, so that one could measure initial rates with different amounts of added product. For each product concentration, the initial velocity for different substrate concentrations would obey the Michaelis–Menten equation, with $V = V^f$ and $K_m = K_m^A(1 + p/K_s^P)$. V is thus independent of p, but K_m increases linearly with p. In fact, product inhibition is sometimes of this type (e.g. the inhibition of invertase by fructose), but sometimes it is not [e.g. the inhibition of invertase by its other product, glucose (Michaelis and Pechstein, 1914)]. Moreover, there are many compounds other than products that inhibit enzymes. It is clear, then, that a more complete theory is required in order to account for these facts, which is developed in later chapters.

Appendix 2.1 Hyperbolic nature of the Michaelis–Menten equation

The plot of v against s according to the Michaelis–Menten equation is often described as a rectangular hyperbola. However, this description sometimes causes perplexity, because the hyperbolas encountered in mathematics always have two limbs, whereas the plot of v against s appears to have only one, and because the Michaelis–Menten equation, $v = Vs/(K_m + s)$, seems to have little in common with the usual expression of a rectangular hyperbola with the x and y axes as asymptotes:

$$xy = a \qquad (2.23)$$

and even less with the alternative expression of a rectangular hyperbola with asymptotes inclined at $45°$ to the x and y axes:

$$x^2 - y^2 = a^2$$

However, substitution of $x = s + K_m$, i.e. $x = s - (-K_m)$, $y = v - V$, $a = -VK_m$ into equation 2.23 gives

$$(s + K_m)(v - V) = -VK_m$$

which is simply a rearranged form of the Michaelis–Menten equation. As the asymptotes of the hyperbola $xy = a$ are the axes $x = 0$ and $y = 0$, it follows from the definitions of x and y that the asymptotes of the plot of v against s are $s = -K_m$ and $v = V$. As the vertical asymptote occurs at a negative value of s, it is now clear why the curve appears to have only one limb: the whole of the negative limb, and part of the positive limb, occur in a physically impossible region of the plot, and so cannot be observed. These conclusions are illustrated in *Figure 2.8*, which shows much more of the curve than it is

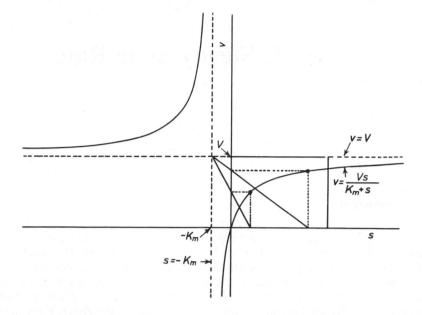

Figure 2.8 Plot of v against s according to the Michaelis–Menten equation: The part of the curve from s = 0 to 5K$_m$ is the same as in Figure 2.1, but a much wider range of values is shown, including physically impossible values, in order to display the relationship of the curve to its asymptotes s = −K$_m$ and v = V

ever possible to observe. This is of practical importance because, whenever one estimates the values of K_m and V from observations of s and v, one is, in effect, estimating the form of the whole hyperbola from a few observations along a short arc.

Figure 2.8 also illustrates a property of the curve pointed out by de Miguel Merino (1974): any straight line through the point of intersection of the asymptotes cuts the axes at two points that together provide the co-ordinates of a point on the curve. This property forms the basis of the direct linear plot, which is, however, drawn as a reflection of *Figure 2.8* about the vertical axis, because it is treated as a plot of V against K_m rather than as a construction on a plot of v against s.

33

3

How to Derive Steady-State Rate Equations

3.1 Introduction

In principle, the steady-state rate equation for any enzyme mechanism can be derived in the same way as that for the simple Michaelis–Menten mechanism: we write down expressions for the rates of change of concentrations of all of the intermediates, set them all equal to zero and solve the simultaneous equations that result. In practice, this method is extremely laborious and liable to error for all but the simplest mechanisms. Fortunately, King and Altman (1956) have described a schematic method that is simple to apply to any mechanism that consists of a series of reactions between different forms of one enzyme. It is not applicable to non-enzymic reactions, to mixtures of enzymes, or to reactions that contain non-enzymic steps. Nonetheless, it is applicable to most of the situations met in enzyme catalysis and is very useful in practice. It is described and discussed in this chapter.

It is not necessary to understand the theory of the King–Altman method in order to apply it, and indeed the theory is considerably more difficult than the practice. Some readers may therefore prefer to proceed directly to the description in Section 3.3. However, it is generally wise to understand a method that one uses, in order to have an understanding of its limitations, and for this reason the theory of the King–Altman method is given in the next section.

3.2 Principle of the King–Altman method

Consider a mechanism in which there are n different enzyme forms, E_1, $E_2, \ldots, E_n$. Suppose that reversible first-order reactions are possible between every pair of species, $E_i \rightleftharpoons E_j$, and let the rate constant for $E_i \rightarrow E_j$ be k_{ij}, and that for $E_j \rightarrow E_i$ be k_{ji}, etc. Then, the rate of production of any particular species E_i is $k_{1i}e_1 + k_{2i}e_2 + \cdots + k_{ni}e_n$, where the summation includes the concentration of every species except E_i itself; and the rate of destruction of E_i is $(k_{i1} + k_{i2} + \cdots + k_{in})e_i$, which we shall represent as $\sum k_{ij}e_i$. Then, the rate of change of e_i is

$$\frac{de_i}{dt} = k_{1i}e_1 + k_{2i}e_2 + \cdots - \sum k_{ij}e_i + \cdots + k_{ni}e_n = 0$$

This expression is equal to zero by virtue of the steady-state assumption. There are n expressions of this type, one for each of the n species. However, only $(n-1)$ of these equations are independent, because any one of them can be obtained by adding the other $(n-1)$ equations together. In order to solve the equations for the n unknowns, it is necessary to have one further equation, which is provided by the condition that the sum of concentrations of all of the species must be e_0, the total enzyme concentration,

$$e_1 + e_2 + \cdots + e_n = e_0 \tag{3.1}$$

It does not matter which of the original n equations is replaced with equation 3.1, but it is convenient when solving for e_m to replace the mth equation. Then we will have a set of simultaneous equations as follows:

$$-\sum k_{1j}e_1 + k_{21}e_2 + \cdots + k_{m1}e_m + \cdots + k_{n1}e_n = 0$$
$$k_{12}e_1 - \sum k_{2j}e_2 + \cdots + k_{m2}e_m + \cdots + k_{n2}e_n = 0$$
$$\vdots$$
$$e_1 + e_2 + \cdots + e_m + \cdots + e_n = e_0$$
$$\vdots$$
$$k_{1n}e_1 + k_{2n}e_2 + \cdots + k_{mn}e_m + \cdots - \sum k_{nj}e_n = 0$$

These n simultaneous equations can now be solved in principle by Cramer's rule, giving, for e_m

$$e_m = \frac{\begin{vmatrix} -\sum k_{1j} & k_{21} & \cdots & 0 & \cdots & k_{n1} \\ k_{12} & -\sum k_{2j} & \cdots & 0 & \cdots & k_{n2} \\ \vdots & \vdots & & \vdots & & \vdots \\ 1 & 1 & \cdots & e_0 & \cdots & 1 \\ \vdots & \vdots & & \vdots & & \vdots \\ k_{1n} & k_{2n} & \cdots & 0 & \cdots & -\sum k_{nj} \end{vmatrix}}{\begin{vmatrix} -\sum k_{1j} & k_{21} & \cdots & k_{m1} & \cdots & k_{n1} \\ k_{12} & -\sum k_{2j} & \cdots & k_{m2} & \cdots & k_{n2} \\ \vdots & \vdots & & \vdots & & \vdots \\ 1 & 1 & \cdots & 1 & \cdots & 1 \\ \vdots & \vdots & & \vdots & & \vdots \\ k_{1n} & k_{2n} & \cdots & k_{mn} & \cdots & -\sum k_{nj} \end{vmatrix}} \tag{3.2}$$

[In this discussion, familiarity with the properties of determinants and the determinant method of solving simultaneous equations ('Cramer's rule') is assumed. To describe these would be beyond the scope of this book, but an account will be found in any good algebra textbook.] Inspection of the numerator of this expression shows that the mth column consists entirely of zeros apart from e_0 in the mth row. This element can be brought into the first row and first column by m switches of rows and m switches of columns, leaving the rest of the determinant unchanged; $2m$ must be even whether m is odd or even, and so the sign of the determinant will be unchanged. As the first column will now consist of zeros apart from e_0 in the first row, it follows that e_0 can be taken out as a factor of the determinant, leaving the order $(n-1)$ determinant, so that the numerator can be written

$$\mathscr{N}_m = e_0 \begin{vmatrix} -\sum k_{1j} & k_{21} & \cdots & k_{n1} \\ k_{12} & -\sum k_{2j} & \cdots & k_{n2} \\ \vdots & \vdots & & \vdots \\ k_{1n} & k_{2n} & \cdots & -\sum k_{nj} \end{vmatrix}$$

If we now examine this determinant, we find:
(1) It contains no constants k_{mj} with m as first index. Therefore, its expansion cannot anywhere contain a constant k_{mj} with m as first index.
(2) Every constant with the same first index occurs in the same column. As every product of constants in the expansion must contain a term from each column, it follows that no product can contain two or more constants with the same first index, and every index other than m must occur once as first index in every product.
(3) Every constant k_{ij}, where $i \neq m$ and $j \neq m$, occurs twice in the determinant, once as a non-diagonal element and once as one of the terms in a $-\sum k_{ij}$ summation. This fact has the very important consequence that every product containing a cycle of indices, such as $k_{12}k_{23}k_{31}$, which contains the cycle $1 \rightarrow 2 \rightarrow 3 \rightarrow 1$, must cancel out in the expansion of the determinant (notice that within a cycle, each of a set of indices occurs once as a first index and once as a second index). In order to see why this should be so, it is simplest to look at a specific example, such as $k_{12}k_{23}k_{31}$. This product will occur as $(-k_{12})(-k_{23})(-k_{31})$, as part of $(-\sum k_{1j})(-\sum k_{2j})(-\sum k_{3j})$, which will be multiplied by terms from the other rows and columns of the determinant, but it will also occur as $+k_{12}k_{23}k_{31}$, from the non-diagonal elements, multiplied by the same terms from the other rows and columns of the determinant. The initial sign is positive, because an even number of switches of columns (switch columns 1 and 2, then 1 and 3) are required to bring these elements on to the main diagonal. Thus, for every product that contains $-k_{12}k_{23}k_{31}$ from the diagonal elements, there will be an equal and opposite product that contains $+k_{12}k_{23}k_{31}$ from the non-diagonal

elements. In general, if there is an odd number of constants in the cycle, as in this example, an even number of switches will be required in order to bring the non-diagonal elements on to the main diagonal, so that the sign will be positive, but the corresponding product from the main diagonal will contain an odd number of negative terms, and so will be negative itself. On the other hand, if there is an even number of constants in the cycle, the opposite will be true. However, in either event the products will cancel out, so that we can generalize and state that *all* cycles cancel from the final expansion.

(4) Any product containing a non-diagonal element must contain at least one other non-diagonal element, because selection of any non-diagonal element removes *two* diagonal elements from the choice of elements available for the rest of the product (for example, if the third element of the fourth row is included, both the third and the fourth diagonal elements are excluded by the requirement that each product must contain one element only from each row and one element only from each column). Then selection of non-diagonal elements can be terminated only by selecting one with first and second indices that have been used already as second and first indices; in other words, a cycle must be completed. However, we have seen that all products that contain cycles must cancel out. Consequently, all products that appear in the final expansion must be derived solely from diagonal elements. As all of the constants in the diagonal are negative, it follows that all of the products in the expansion must have the same sign [positive if $(n-1)$ is even, negative if $(n-1)$ is odd].

(5) We have seen, under point (2), that every index except m must occur at least once as a first index. Each product contains $(n-1)$ constants and so m must occur at least once as a second index because, if it did not, every index that occurred as a second index would also occur as a first index, and the product would inevitably contain at least one cycle.

(6) Every diagonal element $-\sum k_{ij}$ contains every possible second index except i. Consequently, every product that is not forbidden by the preceding rules must appear in the final expansion.

We can summarize the conclusions from the above discussion as follows. The expansion of the numerator of equation 3.2 contains a sum of products of $(n-1)$ k_{ij} constants, in each of which (1) m does not occur as first index, (2) every other index occurs once only as first index, (3) no cycles of indices occur, (4) every product has the same sign, (5) m occurs at least once as second index and (6) every allowed product occurs.

After this lengthy discussion of the numerator of equation 3.2, it will be a relief to discover that it is not necessary to discuss the denominator in equal detail: the denominator has the same value for every enzyme species and, because the total concentration of all the species must be e_0, the denominator must be the sum of all the different numerators, divided by e_0. Because of this, and because all of the possible numerators have the same sign, the denominator must have the same sign also, so that the fraction as a whole must be positive. This, of course, merely confirms the physical necessity that all concentrations be positive.

We have hitherto made the assumption that reactions exist between every pair of species. This is, of course, unrealistic, but the absence of certain reactions can readily be dealt with by assigning zero values to the rate constants. Hence products containing such constants will be zero, and can therefore be omitted.

Another objection to the above discussion is that two or more parallel reactions between two species may exist. In this case, the total forward rate will be the sum of the individual rates, and similarly for the total back rate. So, in the above discussion, any k_{ij} can be considered to be the sum of a number of constants for parallel reactions.

All of the products of rate constants discussed in this section can be regarded as 'trees' or pathways leading to one particular species from each of the other species. Consequently, the method to be described in the next section follows directly from this discussion.

3.3 Method of King and Altman

The following schematic method can be used to derive the rate equation for any mechanism. We shall use for an example one of the most important two-substrate mechanisms:

$$E + A \underset{k_{-1}}{\overset{k_{+1}}{\rightleftharpoons}} EA$$

$$EA + B \underset{k_{-2}}{\overset{k_{+2}}{\rightleftharpoons}} EAB$$

$$EAB \rightleftharpoons EPQ$$

$$EPQ \underset{k_{-3}}{\overset{k_{+3}}{\rightleftharpoons}} EQ + P$$

$$EQ \underset{k_{-4}}{\overset{k_{+4}}{\rightleftharpoons}} E + Q$$

No rate constants are shown for the third reaction, because steady-state measurements provide no information about isomerizations between intermediates that do not involve the take-up or release of reactants, as is discussed in Section 3.7. For analytical purposes, therefore, we must treat EAB and EPQ as a single species, even though it may be mechanistically more meaningful to regard them as distinct.

The first step in the King–Altman method is to represent the mechanism by a scheme that shows all of the enzyme species and the reactions between them:

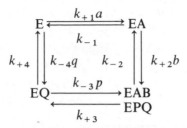

All of the reactions must be treated as first-order reactions; for example, the second-order rate constant k_{+1} is replaced with the pseudo-first-order rate constant $k_{+1}a$ by including the concentration of A.

Next, a *master pattern* is drawn representing the skeleton of the scheme, in this case a square:

It is then necessary to find every pattern that (1) consists only of lines from the master pattern, (2) connects every·enzyme species and (3) contains no closed loops. Each will contain one line fewer than the number of enzyme species, and in this case there are four such patterns:

For each enzyme species, and each pattern, the product of the rate constants in the pattern leading to that species is written down. For example, for EQ, the second pattern gives $k_{-1}k_{+3}k_{-4}q$, from

$$\begin{array}{c} \xleftarrow{k_{-1}} \\ k_{-4}q \\ \xleftarrow{k_{+3}} \end{array}$$

The arrow-heads are drawn (or imagined) so that from any starting point the arrows lead to the appropriate species, EQ, with one arrow only *from* each other species. This ensures that rules 1, 2 and 5 of the previous section are obeyed. Obedience to the other rules is assured by selecting patterns as described above. Since there are four patterns, four such products can be written down for each species. Then, the fraction of each species in the steady-state mixture is the sum of its four products divided by the sum of all 16 products:

39

$$[E]/e_0 = (k_{-1}k_{-2}k_{+4} + k_{-1}k_{+3}k_{+4}$$
$$+ k_{+2}k_{+3}k_{+4}b + k_{-1}k_{-2}k_{-3}p)/\Sigma \qquad (3.3)$$

$$[EA]/e_0 = (k_{+1}k_{-2}k_{+4}a + k_{+1}k_{+3}k_{+4}a$$
$$+ k_{-2}k_{-3}k_{-4}pq + k_{+1}k_{-2}k_{-3}ap)/\Sigma$$

$$[EAB]/e_0 = (k_{+1}k_{+2}k_{+4}ab + k_{-1}k_{-3}k_{-4}pq$$
$$+ k_{+2}k_{-3}k_{-4}bpq + k_{+1}k_{+2}k_{-3}abp)/\Sigma$$

$$[EQ]/e_0 = (k_{-1}k_{-2}k_{-4}q + k_{-1}k_{+3}k_{-4}q$$
$$+ k_{+2}k_{+3}k_{-4}bq + k_{+1}k_{+2}k_{+3}ab)/\Sigma$$

In each case, the resulting expression, divided by the sum of all of the expressions, shown above as Σ, represents the fraction of the total enzyme in each form.

The rate of the reaction is then the sum of the rates of the steps that generate one particular product, minus the sum of the rates of the steps that consume the same product. In this example, there is one step that generates P, (EAB–EPQ) $\xrightarrow{k_{+3}}$ EQ, and one step that consumes P, EQ $\xrightarrow{k_{-3}p}$ (EAB–EPQ), so we have:

$$v = \frac{dp}{dt} = k_{+3}[EAB] - k_{-3}[EQ]p$$
$$= e_0(k_{+1}k_{+2}k_{+3}k_{+4}ab + k_{-1}k_{+3}k_{-3}k_{-4}pq + k_{+2}k_{+3}k_{-3}k_{-4}bpq$$
$$+ k_{+1}k_{+2}k_{+3}k_{-3}abp - k_{-1}k_{-2}k_{-3}k_{-4}pq - k_{-1}k_{+3}k_{-3}k_{-4}pq$$
$$- k_{+2}k_{+3}k_{-3}k_{-4}bpq - k_{+1}k_{+2}k_{+3}k_{-3}abp)/\Sigma$$
$$= e_0(k_{+1}k_{+2}k_{+3}k_{+4}ab - k_{-1}k_{-2}k_{-3}k_{-4}pq)/\Sigma$$

As it is not normally possible to measure all of the separate rate constants, it is convenient to express the equation in *coefficient form*, which permits a straightforward prediction of the experimental properties of a given mechanism:

$$v = \frac{e_0(c_1 ab - c_2 pq)}{c_3 + c_4 a + c_5 b + c_6 p + c_7 q + c_8 ab + c_9 ap + c_{10}bq + c_{11}pq + c_{12}abp + c_{13}bpq}$$

where

$$c_1 = k_{+1}k_{+2}k_{+3}k_{+4}$$
$$c_2 = k_{-1}k_{-2}k_{-3}k_{-4}$$
$$c_3 = k_{-1}(k_{-2} + k_{+3})k_{+4}$$
$$c_4 = k_{+1}(k_{-2} + k_{+3})k_{+4}$$
$$c_5 = k_{+2}k_{+3}k_{+4}$$
$$c_6 = k_{-1}k_{-2}k_{-3}$$
$$c_7 = k_{-1}(k_{-2} + k_{+3})k_{-4}$$
$$c_8 = k_{+1}k_{+2}(k_{+3} + k_{+4})$$
$$c_9 = k_{+1}k_{-2}k_{-3}$$
$$c_{10} = k_{+2}k_{+3}k_{-4}$$

$$c_{11} = (k_{-1} + k_{-2})k_{-3}k_{-4}$$
$$c_{12} = k_{+1}k_{+2}k_{-3}$$
$$c_{13} = k_{+2}k_{-3}k_{-4}$$

At the steady state, the concentrations of all enzyme forms are constant. Plainly, therefore, Q must be produced at the same rate as P, and A and B must each be consumed at the same rate. Thus it does not matter which reactant is considered in writing down the rate equation. It is simple and instructive to confirm that the expressions for dq/dt, $-da/dt$ and $-db/dt$ are all identical with the expression for dp/dt that we have derived.

3.4 Modifications to the King–Altman method

The method of King and Altman as described is convenient and simple to apply to any of the simpler enzyme mechanisms. However, complex mechanisms often require very large numbers of patterns to be found. The derivation is then very laborious, and liable to errors on account of overlooking patterns or writing down incorrect terms. Although it is possible in principle to calculate the total number of patterns, it is very tedious unless the mechanism is very simple, because corrections must be applied for all cycles of reactions within the mechanism. In any event, knowing the number of patterns to be found may not be very helpful in finding them, and does not reduce the labour involved in writing down the terms. In general, for complex mechanisms, it is better to search for means of simplifying the procedure. A number of rules for carrying out this exercise have been given by Volkenstein and Goldstein (1966), using the theory of flow graphs, which has developed by Mason (1953, 1956) for the analysis of electronic networks. The simplest of these rules are as follows:
(1) If there are two or more steps interconverting the same pair of enzyme forms, these steps can be condensed into one by adding the rate constants of the parallel reactions. For example, the Michaelis–Menten mechanism is represented in the King–Altman method as

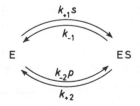

which gives the two patterns ⌢ and ⌣. Because the two reactions connect the same pair of enzyme forms, they can be added, to give

$$E \underset{k_{-1}+k_{+2}}{\overset{k_{+1}s + k_{-2}p}{\rightleftharpoons}} ES$$

41

This scheme is itself the only pattern, so that

$$[E]/e_0 = (k_{-1}+k_{+2})/(k_{-1}+k_{+2}+k_{+1}s+k_{-2}p)$$
$$[ES]/e_0 = (k_{+1}s+k_{-2}p)/(k_{-1}+k_{+2}+k_{+1}s+k_{-2}p)$$

In more complex cases, the simplification afforded by this technique is very great: an example discussed by King and Altman was the general modifier mechanism of Botts and Morales (1953):

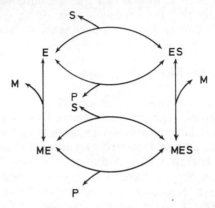

As shown, this master pattern requires twelve patterns, but if the parallel paths are added a square is obtained, which requires only four patterns.

(2) If the mechanism contains different enzyme forms that have identical properties, the procedure is greatly simplified by treating such forms as single species. For example, if an enzyme contains two *identical* active sites, the mechanism might be represented by

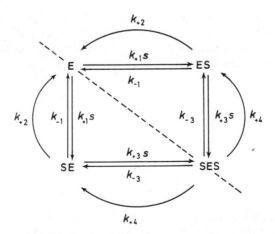

which requires 32 patterns. However, as ES and SE are identical, the pattern is symmetrical about the broken line, and can be represented much more simply as

42

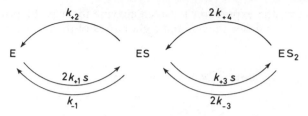

which can be further simplified by rule 1 to

$$\text{E} \xrightleftharpoons[k_{-1}+k_{+2}]{2k_{+1}s} \text{ES} \xrightleftharpoons[2(k_{-3}+k_{+4})]{k_{+3}s} \text{ES}_2$$

Thus, a scheme of 32 patterns has been simplified to one of a single pattern, and the expressions for the three species can be written down immediately:

$$[\text{E}]/e_0 = 2(k_{-1}+k_{+2})(k_{-3}+k_{+4})/[2(k_{-1}+k_{+2})(k_{-3}+k_{+4}) \\ + 4k_{+1}(k_{-3}+k_{+4})s+2k_{+1}k_{+3}s^2]$$

and so on.

Whenever advantage is taken of the symmetry of the master pattern in this way in order to condense it into a simpler scheme, statistical factors appear. In this example, the reaction $\text{E} \to \text{ES}$ can occur in two ways, so that the total rate is the sum of the two rates, giving a rate constant that is double the rate constant for either of the two paths. The back reaction, in contrast, can occur in only one way, and so has a statistical factor of 1. Therefore

$$
\begin{array}{c}
\text{E} \xrightleftharpoons[k_{-1}]{k_{+1}s} \text{ES} \\[4pt]
k_{-1} \Big\Updownarrow k_{+1}s \\[4pt]
\text{SE}
\end{array}
$$

becomes

$$\text{E} \xrightleftharpoons[k_{-1}]{2k_{+1}s} \text{ES}$$

(3) If the master pattern consists of two or more distinct parts touching at single enzyme forms, it is convenient to treat the different parts separately. A simple example of this is provided by the case of *competitive substrates*, where a single enzyme catalyses two reactions with different substrates simultaneously:

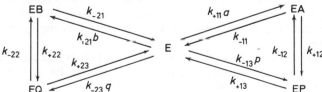

In this case, the expression for each enzyme form is the product of the appropriate sums for the left and right halves of the master pattern:

$$[E]/e_0 = (k_{+22}k_{+23}+k_{-21}k_{-22}+k_{-21}k_{+23})$$
$$(k_{+12}k_{+13}+k_{-11}k_{-12}+k_{-11}k_{+13})/\Sigma$$

$$[EA]/e_0 = (k_{+22}k_{+23}+k_{-21}k_{-22}+k_{-21}k_{+23})$$
$$(k_{-12}k_{-13}p+k_{+11}k_{-12}a+k_{+11}k_{+13}a)/\Sigma$$

$$[EP]/e_0 = (k_{+22}k_{+23}+k_{-21}k_{-22}+k_{-21}k_{+23})$$
$$(k_{+12}k_{-13}p+k_{+11}k_{+12}a+k_{-11}k_{-13}p)/\Sigma$$

$$[EB]/e_0 = (k_{-22}k_{-23}q+k_{+21}k_{-22}b+k_{+21}k_{+23}b)$$
$$(k_{+12}k_{+13}+k_{-11}k_{-12}+k_{-11}k_{+13})/\Sigma$$

$$[EQ]/e_0 = (k_{+22}k_{-23}q+k_{+21}k_{+22}b+k_{-21}k_{-23}q)$$
$$(k_{+12}k_{+13}+k_{-11}k_{-12}+k_{-11}k_{+13})/\Sigma$$

This last modification can be regarded as a special case of the procedure that we shall describe in the next section, but because it is very easy to apply it is convenient to treat it separately.

Various other methods of deriving rate equations have been described that are claimed to be superior to the method of King and Altman. Of these, the method of Fromm (1970) may prove useful to those who prefer a more algebraic and less geometric approach. Needless to say, all valid methods should give equivalent rate equations.

3.5 Compression of patterns

The fourth modification to the King–Altman method introduced by Volkenstein and Goldstein (actually the third in their enumeration) is the most useful and important, because it provides the only practical method of analysing complex mechanisms with six or more enzyme forms. Unfortunately, it is also the most difficult to understand and use, because it is not purely mechanical but requires careful thought if it is to be used profitably. In essence, it provides a means of recognizing and using repetitive features of the master pattern, so that one can write down terms for several patterns simultaneously.

Let us consider the following mechanism for a two-substrate, two-product reaction in which the substrates can bind in either order and the products can be released in either order:

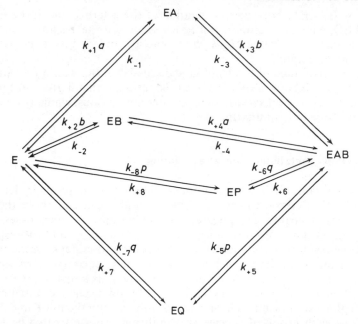

It is scarcely practicable to apply the full King–Altman method to such a complex mechanism, because 32 patterns are required and the probability of errors is great.

If we consider pathways that terminate at EAB, it is clear from inspection that every valid pattern must contain five lines and must include one and only one of the fragments $E \to EA \to EAB$ *or* $E \to EB \to EAB$ *or* $E \to EP \to EAB$ *or* $E \to EQ \to EAB$. Any pattern containing $E \to EA \to EAB$ must also contain *either* $EB \to E$ *or* $EB \to EAB$, but not both, in order for there to be a connection to EB and no closed loop. The composite term $(k_{-2} + k_{+4}a)$ must therefore be a factor of the sum of terms for the patterns that contain $E \to EA \to EAB$. As every pattern must also contain connections to EP and EQ, the same argument applies to them also, and $(k_{-6}q + k_{+8})$ and $(k_{-5}p + k_{+7})$ must be factors of the same sum. Consequently, we can write down a single term, $k_{+1}k_{+3}ab(k_{-2} + k_{+4}a)(k_{-6}q + k_{+8})(k_{-5}p + k_{+7})$, that accounts for all of the eight patterns that contain the fragment $E \to EA \to EAB$. The other three two-step fragments can be treated in a similar manner, so that the 32 terms in the equation for EAB can be expressed by the sum of four factored terms.

The terms in the equation for E can be found by simply transposing those for EAB. Thus, $E \to EA \to EAB$ becomes $E \leftarrow EA \leftarrow EAB$, and so in the first sum $k_{+1}k_{+3}ab$ is replaced with $k_{-1}k_{-3}$, and so on.

The terms for EA require a little more thought, because this species occupies a position in the master pattern that is topologically dissimilar from those occupied by E and EAB. For example, the fragment $E - EB - EAB$ is $E \to EB \to EAB$, represented by $k_{+2}k_{+4}ab$, if it is connected to EA by $EAB \to EA$, but it is $E \leftarrow EB \leftarrow EAB$, represented by $k_{-2}k_{-4}$, if it is con-

45

nected by $E \rightarrow EA$. However, evaluation of these terms, and those for EB, EP and EQ, which are similar, will be left as an exercise, because proficiency with this technique comes only with practice and further explanation is unlikely to be helpful.

A major advantage of the technique discussed in this section, in addition to reducing the labour of deriving a rate equation, is that it gives an equation in which repetitive and symmetrical features are obvious, which is very helpful for locating and eliminating errors.

3.6 Reactions containing steps at equilibrium

Some mechanisms are important enough to be worth analysing in detail, but so complex that even with the aid of the methods described above they give rise to unmanageably complicated rate equations. In such cases, some simplifying assumptions are unavoidable and great simplifications often result if one assumes that some steps, such as protonation steps, are maintained at equilibrium at all times. Such assumptions may, of course, turn out to be false after further investigation, but they are useful as a first approximation.

Cha (1968) has described a method for analysing mechanisms that contain some steps at equilibrium, which is much simpler than the full King–Altman analysis as each group of species at equilibrium can be treated as a single species. As an example, suppose we have two species X and Y at equilibrium with one another such that $[Y]/[X] = K$, and that X can be interconverted in a slow reaction with a third species Z:

$$- - - \; X \underset{k_{-1}}{\overset{k_{+1}}{\rightleftharpoons}} Z \; - - -$$
$$\Big\updownarrow K$$
$$- - - \; Y$$

(The broken lines are included in the diagram to emphasize that this is a fragment of a complex mechanism and not a complete mechanism.) The rate of the slow reaction in the forward direction is $k_{+1}[X]$, but this can be written equally well as $k_{+1}[X]([X]+[Y])/([X]+[Y]) = k_{+1}([X]+[Y])/(1+K)$. In other words, X and Y can be treated as a single species, with concentration $([X]+[Y])$, and the rate constant k_{+1} for the breakdown of X is reduced to $k_{+1}/(1+K)$ for the breakdown of the composite species. In general, any number of species in equilibrium can be treated as one species, and each rate constant k_i is reduced to $f_i k_i$, where f_i is the fraction of reactive molecules in the equilibrium mixture. If more than one species is reactive, the rate constant is the sum of the reduced rate constants for the parallel reactions, in accordance with rule 1 in Section 3.4. Therefore, in the example given above, if Y could also break down to Z, or to a species in equilibrium with Z, with a rate constant k_{+2}, the rate constant for the breakdown of the composite species would be $(k_{+1}+k_{+2}K)/(1+K)$.

This type of simplification is particularly useful in the analysis of pH

46

dependence (Chapter 7) and in the analysis of mechanisms with parallel pathways. In the latter case, it is often convenient to treat the alternative pathways as equilibria and the compulsory pathways as slow steps. Equations derived in this way are generally in accordance with experiment, but *this does not mean that the underlying assumptions are correct*: Gulbinsky and Cleland (1968) have shown by computer simulation that it is possible and indeed likely that the additional terms in the rigorous steady-state equation may be numerically significant and yet virtually impossible to detect, because of near proportionality to other terms in the equation over any reasonable experimental range.

3.7 Analysing mechanisms by inspection

The compression of patterns described in the previous section is an important example of the use of inspection for analysing mechanisms: once one is thoroughly conversant with the King–Altman method, it is often possible to reach important conclusions about the rate equation for a mechanism without having to derive it in detail, simply by inspecting the master pattern carefully.

It is an important characteristic of the King–Altman method that every pattern generates a positive term, and that every term appears in the denominator of the rate equation. As there are no negative terms, no terms can cancel by subtraction, and so every term for which a pattern exists must appear in the rate equation. The only exception to this rule is that sometimes the numerator and denominator share a common factor that can be cancelled by division, which normally happens only if the rate constants are related to one another, as in the following mechanism for an enzyme with two independent and identical active sites:

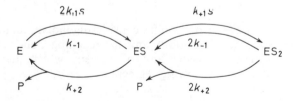

In this case, the rate constants for the second site are the same as those for the first, apart from statistical factors, and the rate equation is

$$v = \frac{4k_{+1}k_{+2}e_0s(k_{-1}+k_{+2})+4k_{+1}k_{+2}e_0s^2}{2(k_{-1}+k_{+2})^2+4k_{+1}s(k_{-1}+k_{+2})+2k_{+1}^2s^2}$$

This equation apparently contains terms in s^2, but actually the numerator and denominator share a common factor, $2(k_{-1}+k_{+2}+k_{+1}s)$, and the rate equation simplifies to the Michaelis–Menten equation:

47

$$v = \frac{2k_{+2}e_0 s}{\left(\dfrac{k_{-1}+k_{+2}}{k_{+1}}\right) + s}$$

In mechanisms where there are no relationships between the rate constants other than those required by thermodynamics, it is safe to assume that cancellation between numerator and denominator will not be possible, so that any term for which a pattern exists must appear in the rate equation. For example, consider the general modifier mechanism of Botts and Morales (1953):

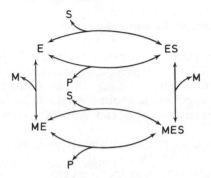

If one wished to confirm that the rate equation for this mechanism contained terms in s^2, one could do so without deriving it by noting that there are two patterns that give rise to terms in s^2:

Apart from the value of inspection for considering individual mechanisms, it can be used to reach important and far-reaching conclusions about mechanisms in general. Consider any mechanism, which may be highly complex, that contains the following segment:

where X, Y and Z are three enzyme species. The rate equation will contain terms involving k_{yx}, terms involving k_{yz} and terms involving neither, but never both, so that $k_{yx}k_{yz}$ will never appear as a product. Suppose now that Y is replaced with a mixture of two interconvertible species, Y_1 and Y_2:

$$X \underset{k_{yx}}{\overset{k_{xy}}{\rightleftharpoons}} Y_1 \underset{k_{21}}{\overset{k_{12}}{\rightleftharpoons}} Y_2 \underset{k_{zy}}{\overset{k_{yz}}{\rightleftharpoons}} Z$$

How does this affect the rate equation? Firstly, any product that appeared in the original rate equation will reappear, multiplied by k_{12} or by k_{21}, so that no combinations of reactant concentrations that were originally present will disappear. However, in addition, there will be products that contain neither k_{12} nor k_{21}, but contain *both* k_{yx} and k_{yz}. If either of these is a simple first-order rate constant, the form of the rate equation will not be affected, but if both k_{yx} and k_{yz} are associated with reactant concentrations, e.g. $k_{yx}a$ and $k_{yz}b$, then terms that contain ab that were absent from the rate equation for the original mechanism appear in the rate equation for the modified mechanism. The important conclusion to be drawn from this discussion is that isomerization of an enzyme species has no effect on the *form* of the rate equation unless reactants can bind to both isomers. This conclusion means that most types of isomerization cannot be detected by steady-state kinetics (although they can be detected by transient-state kinetics: *see* Chapter 9) and that the rate constants that appear in the rate equation may actually be combinations of rate constants for several elementary steps of the mechanism. As a 'rule of thumb,' it is usual to distinguish between isomerization of the free enzyme, which can be detected in principle by steady-state measurements, and isomerization of transitory complexes, which usually cannot. Actually, this discussion may be academic, because steady-state measurements have never successfully identified an isomerization, either of the free enzyme or of another species. In fact, in at least one instance, an isomerization has been shown to occur by other methods, but has failed to manifest itself in steady-state rate measurements. This aspect is discussed further in Section 5.10.

The reason why reversible steps in a mechanism cannot be treated as equilibria is that any net flux through a step must have the effect of unbalancing any equilibrium that might otherwise exist. However, in a *dead-end reaction*, i.e. one connected to the rest of the mechanism at one end only, there is no net flux, and so there is no reason why equilibrium should not be maintained. Consider, for example, the mechanism discussed in Section 3.3, with the addition of a dead-end complex, EBQ:

$$
\begin{array}{ccc}
E & \underset{k_{-1}}{\overset{k_{+1}a}{\rightleftharpoons}} & EA \\
k_{+4} \updownarrow k_{-4}q & & k_{-2} \updownarrow k_{+2}b \\
EBQ \underset{k_{+5}b}{\overset{k_{-5}}{\rightleftharpoons}} EQ & \underset{k_{+3}}{\overset{k_{-3}p}{\rightleftharpoons}} & \begin{array}{c} EAB \\ EPQ \end{array}
\end{array}
$$

For every species except EBQ, the King–Altman analysis gives the same

49

expression as before, multiplied by k_{-5}; EBQ requires the original expression for EQ multiplied by k_{+5}. Thus, $[EBQ]/[EQ] = k_{+5}b/k_{-5}$, and EQ, B and EBQ are in equilibrium. The rate equation for the mechanism with dead-end inhibition is therefore identical with the rate equation without such inhibition, except that the terms for EQ in the denominator must be multiplied by $(1 + k_{+5}b/k_{-5})$.

Finally, inspection can be used to establish a general principle about mechanisms that contain two species that cannot react in unimolecular steps. For example, in the following mechanism, E and E′ react only in bimolecular steps:

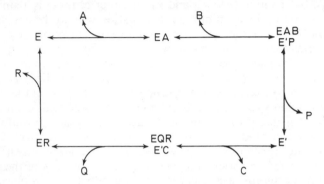

As every King–Altman pattern must contain steps from E or from E′ or both, every term in the rate equation must contain at least one reactant concentration; in other words, the rate equation must lack a constant. By an extension of this argument, if there are three such species, every term in the rate equation must contain at least two reactant concentrations. Mechanisms of this type are called *substituted-enzyme mechanisms*, and are discussed further in Chapter 5.

3.8 Rate equations in coefficient form

It is often convenient to express rate equations in coefficient form, as was done in Section 3.3 (equation 3.4). Firstly, the equation is simpler than that which contains rate constants; secondly, the rate constants may not be true individual rate constants, because there may be isomerizations that have not been detected; and thirdly, the coefficients are in principle measurable quantities. Nonetheless, it is important to be aware of any relationships that may exist between the coefficients. Consider the following two equations:

$$v = \frac{Vs}{K_{m1}+s} + \frac{Vs}{K_{m2}+s} \tag{3.5}$$

$$v = \frac{V[(Ls/K_T)(1+s/K_T)+(s/K_R)(1+s/K_R)]}{L(1+s/K_T)^2+(1+s/K_R)^2} \tag{3.6}$$

in which K_{m1}, K_{m2}, L, K_R and K_T are all positive constants. Both equations apply to enzymes that contain two active sites, but equation 3.5 assumes that these have different K_m values and act independently, whereas equation 3.6 assumes that they are related according to the symmetry model of Monod, Wyman and Changeux (1965; cf. Section 7.7). In both cases, the two sites are assumed to have identical catalytic constants. However, for the present purpose, the meanings of the two equations are not important; the point to note is that in coefficient form the two equations are identical:

$$v = \frac{c_1 s + c_2 s^2}{c_3 + c_4 s + s^2}$$

However, it would be wrong to conclude that they are indistinguishable: the difficulty with this equation is that it suppresses any relationships that may exist between the coefficients. Consider c_3 and c_4: for equation 3.5, they are given by

$$c_3 = K_{m1} K_{m2}$$
$$c_4 = K_{m1} + K_{m2}$$

Because of the requirement that K_{m1} and K_{m2} must both be positive, c_3 and c_4 are not completely independent, but are related by the inequality

$$c_4^2 \geqslant 4c_3 \tag{3.7}$$

For equation 3.6, c_3 and c_4 are given by

$$c_3 = \frac{L+1}{\dfrac{L}{K_T^2} + \dfrac{1}{K_R^2}}$$

$$c_4 = \frac{2\left(\dfrac{L}{K_T} + \dfrac{1}{K_R}\right)}{\dfrac{L}{K_T^2} + \dfrac{1}{K_R^2}}$$

These two equations also imply a relationship between c_3 and c_4, which in this case is

$$c_4^2 \leqslant 4c_3 \tag{3.8}$$

Apart from the special case where $c_4^2 = 4c_3$, equations 3.7 and 3.8 are contradictory. Therefore, the original models defined by equations 3.5 and 3.6 do not overlap, and in any experiment it should be possible to exclude one or other of them.

4

Inhibitors and Activators

4.1 Reversible and irreversible inhibitors

Compounds that reduce the rate of an enzyme-catalysed reaction when they are added to the reaction mixture are called *inhibitors*. Inhibition can arise in a wide variety of ways, however, and there are many different types of inhibitor. A class that is of little importance in enzyme kinetics (except as a nuisance) is that of irreversible inhibitors or catalytic poisons. Inhibitors of this type combine with the enzyme in such a way as to reduce its activity to zero. Many enzymes are poisoned by trace amounts of heavy metal ions, and for this reason it is common practice to carry out kinetic studies in the presence of complexing agents, such as ethylenediamine tetraacetate. This is particularly important in the purification of enzymes: in crude preparations, the total protein concentration is high and the many protein impurities sequester almost all of the metal ions that may be present, but the purer an enzyme becomes, the less it is protected by other proteins and the more important it is to add alternative sequestering agents. Irreversible inhibitors are occasionally used in a positive way. For example, poisoning by mercury-(II) compounds has often been used to implicate sulphydryl groups in the activity of enzymes. However, this application is essentially qualitative and non-kinetic, and catalytic poisons will not be discussed further.

A much more important class of inhibitors is that of reversible inhibitors. These inhibitors form dynamic complexes with the enzyme that have different catalytic properties from those of the free enzyme. The inhibited enzyme may have an increased K_m value (*competitive inhibition*), a reduced V value (*pure non-competitive inhibition*). V and K_m reduced in a constant ratio (*uncompetitive inhibition*), or some combination of these effects (*mixed inhibition*).

As a basis for discussion, it is useful to examine a general scheme proposed by Botts and Morales (1953), which includes most of the simple types of inhibition as special cases:

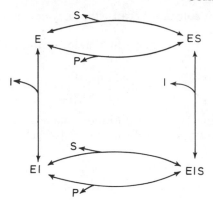

In this scheme, there are four species, E, ES, EI and EIS, with six reactions between them. All of the simple cases can be generated by omitting some of the reactions: thus the inhibition is competitive (Section 4.2) if EIS and the reactions involving it are missing; it is uncompetitive (Section 4.4) if EI is missing; and it is mixed (Section 4.3) if EI and EIS both occur but are not directly interconvertible. The Botts–Morales scheme is most useful for discussing inhibitors that are *not* reaction products: although these form an important class of inhibitors and behave kinetically in a similar manner to non-product inhibitors, their relationship to the scheme is less obvious.

Some workers make a distinction between a species EI, in which the inhibitor is bound to the substrate binding site, and IE, in which it is bound to a different site. This distinction is not particularly helpful and complicates the scheme unnecessarily, because mechanisms do not normally arise in which *both* EI and IE occur. In general, if EIS is a significant species, one can assume that the inhibitor and substrate bind to different sites.

The Botts–Morales scheme also includes certain *activator* mechanisms, in which the added substance increases the reaction rate. It is therefore often referred to as a general *modifier* mechanism, where the term modifier embraces both inhibitors and activators.

4.2 Competitive inhibition

The commonest type of inhibition is termed *competitive inhibition*, because the simplest explanation of it is that the inhibitor binds to the same site on the enzyme as the substrate, forming an abortive (i.e. non-productive) complex. In other words, the substrate and inhibitor compete for the same site, so that only one enzyme–inhibitor complex, EI, is possible. In the simplest type of competitive inhibition, EI is a dead-end complex, as it can break down only by returning to $E + I$. Consequently (cf. Section 3.7), its concentration is given by a true equilibrium constant, $K_i = [E][I]/[EI]$, which is termed the *inhibition constant*. In many of the more complex types of inhibition, including most types of product inhibition, the inhibition constant cannot be treated as a true equilibrium constant because the enzyme–inhibitor complex is not a dead-end complex.

The complete steady-state rate equation for simple competitive inhibition is as follows:

$$v = \frac{Vs}{K_m(1 + i/K_i) + s} \tag{4.1}$$

where i is the free inhibitor concentration and V and K_m have their usual meanings. The equation is of the form of the Michaelis–Menten equation, i.e. it can be written as

$$v = \frac{V^{app}s}{K_m^{app} + s}$$

where V^{app} and K_m^{app} are the 'apparent' values of V and K_m and are given by

$$V^{app} = V$$

$$K_m^{app} = K_m(1 + i/K_i)$$

$$V^{app}/K_m^{app} = \frac{V/K_m}{1 + i/K_i}$$

Hence the effect of a competitive inhibitor is to increase the apparent value of K_m by the factor $(1 + i/K_i)$, to reduce that of V/K_m by the same factor, and to leave V unchanged. V/K_m is mentioned explicitly here because in most situations (although not this one) its behaviour is simpler than that of K_m.

4.3 Mixed inhibition

Most elementary accounts of inhibition discuss two types of inhibition only, competitive inhibition and *non-competitive inhibition*. Competitive inhibition is of genuine importance, but non-competitive inhibition is a phenomenon that does not occur in most practical situations and it need not be considered in detail here. It arose originally because the earliest students of inhibition, Michaelis and his collaborators, assumed that certain inhibitors acted by reducing the apparent value of V, but had no effect on K_m. This effect would be an obvious alternative to competitive inhibition, and was termed non-competitive inhibition. However, it is difficult to imagine a reasonable explanation of such effects: one would have to assume that the inhibitor interfered with the catalytic properties of the enzyme, but that it had no effect on the binding of substrate. This might be possible for very small inhibitors, such as protons or metal ions, but seems most unlikely otherwise. In fact, non-competitive inhibition or activation by protons is common and there are several instances of non-competitive inhibition by heavy-metal ions. However, non-competitive inhibition by other species is very rare, and most of the commonly quoted examples, such as the inhibition of invertase by α-glucose (Nelson and Anderson, 1926) and the inhibition of arginase by various compounds (Hunter and Downs, 1945), prove, on re-examination of the original

54

data, to be examples of mixed inhibition. In general, it is best to regard non-competitive inhibition as a special, and not very interesting, case of *mixed inhibition*, which is discussed below.

Mixed inhibition occurs when both V^{app} and V^{app}/K_m^{app} vary with the inhibitor concentration. In the simplest case the following equations apply:

$$V^{app} = \frac{V}{1 + i/K_i'} \tag{4.2}$$

$$K_m^{app} = \frac{K_m(1 + i/K_i)}{1 + i/K_i'} \tag{4.3}$$

$$V^{app}/K_m^{app} = \frac{V/K_m}{1 + i/K_i} \tag{4.4}$$

This type of inhibition can be accommodated by the Botts–Morales scheme if EIS does not break down to products and if all binding reactions can be treated as equilibria. In this case, K_i is the dissociation constant of EI, K_i' the inhibitor dissociation constant of EIS and K_m the dissociation constant of ES, i.e. K_s. Actually, the deviations from these equations that occur if the binding reactions are not truly equilibria are usually very small, and consequently one cannot use adherence to the equations as evidence that K_m, K_i and K_i' are true dissociation constants.

Mixed inhibition occurs much more commonly as a particular case of product inhibition. If a product is released in a step that generates an enzyme species other than that to which the substrate binds, product inhibition is expected to be in accordance with equations 4.2–4.4. This conclusion is not dependent on any equilibrium assumptions, i.e. it is a necessary consequence of the steady-state treatment, as can readily be shown by the methods of Chapter 3. The simplest of many mechanisms of this type is one in which product is released in the second of three steps:

$$\mathrm{E} + \mathrm{S} \underset{k_{-1}}{\overset{k_{+1}}{\rightleftharpoons}} \mathrm{ES} \underset{k_{-2}}{\overset{k_{+2}}{\rightleftharpoons}} \mathrm{E}' + \mathrm{P}$$
$$k_{+3}$$

More complex examples abound in reactions that involve more than one substrate or product, as will be seen in Chapter 5. In these cases, identification of K_i and K_i' with dissociation constants is not very useful. Even in this simple example, $K_i = (k_{-1} + k_{+2})k_{+3}/k_{-1}k_{-2}$ and $K_i' = (k_{+2} + k_{+3})/k_{-2}$, neither of which is an equilibrium constant except in special cases, such as $k_{+3} \ll k_{+2}$. If $k_{-1} = k_{+3}$, K_i and K_i' are equal, and equation 4.3 simplifies to $K_m^{app} = K_m$. This is, of course, the condition for non-competitive inhibition, but there is no particular reason to expect k_{-1} and k_{+3} to be equal and so there is no particular reason to expect non-competitive inhibition to occur in this, or any other, case of product inhibition.

Because of the rareness of non-competitive inhibition, some enzymologists have generalized the term to include mixed inhibition. There seems to be no advantage in doing this, and it is a most unfortunate development, as it has added ambiguity to an already confused situation. In order to avoid this ambiguity, it is necessary to refer to non-competitive inhibition as *pure* non-competitive inhibition, on the rare occasions when one wishes to refer to it at all.

4.4 Uncompetitive inhibition

The last of the simple types of inhibition to be considered is known, rather unhelpfully, as *uncompetitive* inhibition, and is characterized by equal effects on V and K_m but no effect on V/K_m:

$$V^{app} = \frac{V}{1 + i/K_i'}$$

$$K_m^{app} = \frac{K_m}{1 + i/K_i'}$$

$$V^{app}/K_m^{app} = V/K_m$$

Comparison of these equations with equations 4.2–4.4 shows that uncompetitive inhibition is an asymptotic case of mixed inhibition in which K_i approaches infinity. Hence it is the converse of competitive inhibition, which is the other asymptotic case of mixed inhibition in which K_i' approaches infinity.

Uncompetitive inhibition is predicted for the Botts–Morales mechanism in the special case in which EI is not formed and EIS occurs as a dead-end complex. This implies that the inhibitor-binding site becomes available only after the substrate has bound. This could happen by an induced-fit mechanism (Section 7.6), and appears to be the best explanation of the observation of uncompetitive inhibition of alkaline phosphatase by L-phenylalanine (Ghosh and Fishman, 1966). In other cases, the inhibitor may bind to a site made available by the release of one product, as in the following mechanism, which is known as a *substituted-enzyme mechanism* (Section 5.2):

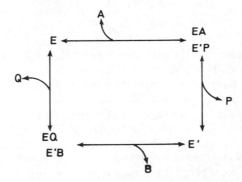

In this mechanism, a dead-end inhibitor that binds only to E′ will be uncompetitive with respect to one substrate, A, whereas an inhibitor that binds only to E will be uncompetitive with respect to the other substrate, B.

56

Uncompetitive inhibition is most common as a case of product inhibition, particularly for reactions with three or more products (Section 5.8). In general, a product acts as an uncompetitive inhibitor if there exists no reversible pathway between the enzyme form to which it binds and the enzyme form to which the substrate binds.

4.5 Plotting inhibition results

All of the types of inhibition that have been described in the three preceding sections are examples of *linear inhibition*, so-called because for all of them $1/V^{app}$ and K_m^{app}/V^{app} display a simple linear dependence on the inhibitor concentration. Linear inhibition is also sometimes termed *complete inhibition*, because the velocity approaches zero if the inhibitor concentration is high enough. Other types of inhibition are possible, as will be seen in Section 4.7, but here we shall confine our attention to linear inhibition.

The properties of linear inhibitors are summarized in *Table 4.1*, where it

Table 4.1 CHARACTERISTICS OF LINEAR INHIBITORS

Type of inhibition	V^{app}	V^{app}/K_m^{app}	K_m^{app}
Competitive	V	$\dfrac{V/K_m}{1+i/K_i}$	$K_m(1+i/K_i)$
Mixed	$\dfrac{V}{1+i/K_i'}$	$\dfrac{V/K_m}{1+i/K_i}$	$\dfrac{K_m(1+i/K_i)}{(1+i/K_i')}$
Uncompetitive	$\dfrac{V}{1+i/K_i'}$	V/K_m	$\dfrac{K_m}{1+i/K_i'}$

can be seen that the effects of inhibitors on V^{app} and V^{app}/K_m^{app} are simple, regular and easily remembered. In contrast, the effects on K_m^{app} are complex and confusing, and so it is advisable for mnemonic purposes to regard K_m^{app} as the ratio of V^{app} and V^{app}/K_m^{app}, rather than as a parameter in its own right. Any of the plots described in Section 2.5 can be used to diagnose the type of inhibition, as they all provide estimates of the apparent values of the kinetic parameters. For example, if plots of s/v against s are made at several values of i, the intercept on the ordinate (K_m^{app}/V^{app}) varies with i if there is a competitive component in the inhibition, and the slope ($1/V^{app}$) varies with i if there is an uncompetitive component. Alternatively, if direct linear plots of V^{app} against K_m^{app} are made at each value of i, the common intersection point shifts in a direction that indicates the type of inhibition: for competitive inhibition, the shift is to the right; for uncompetitive inhibition, it is towards the origin; and for mixed inhibition, it is intermediate between these extremes. These plots are illustrated in *Figure 4.1*.

Other plots are needed for determining the actual values of K_i and K_i'. The simplest approach is to estimate the apparent kinetic constants at several values of i, and to plot K_m^{app}/V^{app} and $1/V^{app}$ against i. In each case, a straight

57

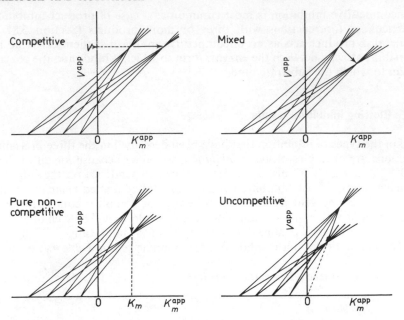

Figure 4.1 Effect of various types of inhibition on the location of the common intersection point (K_m^{app}, V^{app}) *of the direct linear plot (cf. Figure 2.7)*

line is obtained, and the intercept on the i axis gives $-K_i$ if K_m^{app}/V^{app} is plotted or $-K_i'$ if $1/V^{app}$ is plotted. Now, it may seem more natural to determine K_i by plotting K_m^{app} rather than K_m^{app}/V^{app} against i, but this is not advisable for the following two reasons. It is valid only if the inhibition is competitive, and gives a curve rather than a straight line if the inhibition is mixed; it is also much less accurate, even if the inhibition is competitive, because K_m^{app} can never be estimated as precisely as K_m^{app}/V^{app}.

Another method of estimating K_i, introduced by Dixon (1953), is also in common use. If the full equation for mixed inhibition,

$$v = \frac{Vs}{K_m(1+i/K_i)+s(1+i/K_i')}$$ (4.5)

is inverted, we obtain

$$\frac{1}{v} = \frac{(K_m+s)}{Vs} + \frac{(K_m/K_i+s/K_i')i}{Vs}$$

so that a plot of $1/v$ against i is a straight line. If two such lines are drawn at different values of s, the point of intersection can be calculated by equating the two expressions for $1/v$. It is found that the lines intersect at a point where $i = -K_i$. This method provides the value of K_i for any of the linear types of inhibition. In uncompetitive inhibition, K_i is infinite and so the lines are parallel.

Although the Dixon plot does not provide the value of K_i', the uncompetitive
58

inhibition constant, this value can be found by plotting s/v against i at several i values (Cornish-Bowden, 1974). In this case, a different set of straight lines is obtained that intersect at $i = -K_i'$. Both types of plot are illustrated in *Figure 4.2*.

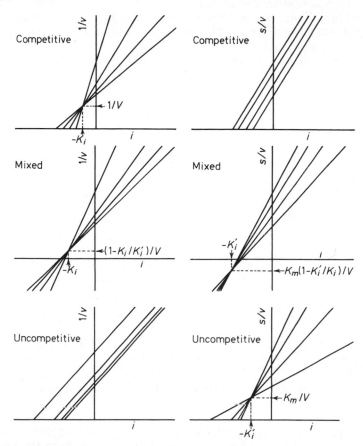

Figure 4.2 Determination of (a) K_i, from plots of $1/v$ against i at various s values, and (b) K_i', from plots of s/v against i at various s values: In the case of mixed inhibition, the point of intersection can be above the axis in the first plot and below it in the second, or vice versa, or, if $K_i = K_i'$ (pure non-competitive inhibition), on the axis in both plots

If one wishes to determine both K_i and K_i' from a single plot, one can do so (in principle) by means of a plot described by Hunter and Downs (1945). It is necessary to know the uninhibited velocity, v_0, at each substrate concentration, as well as v_i, the inhibited velocity, but this does not usually present any problem in practice. Writing v as v_i and $Vs/(K_m+s)$ as v_0, we can rearrange equation 3.5 to give

$$\frac{iv_i}{v_0-v_i} = \frac{K_m+s}{K_m/K_i+s/K_i'} \qquad (4.6)$$

59

In competitive inhibition, $K_i' \to \infty$, and so this equation simplifies to

$$\frac{iv_i}{v_0 - v_i} = K_i(1 + s/K_m)$$

which is the equation for a straight line with an intercept K_i on the ordinate. For mixed inhibition, the line defined by equation 4.7 is a rectangular hyperbola, but it still has an intercept K_i on the ordinate. Instead of increasing to infinity as s is increased, $iv_i/(v_0 - v_i)$ approaches a limiting value of K_i', as illustrated in *Figure 4.3*. In practice, it is likely to be difficult to locate both

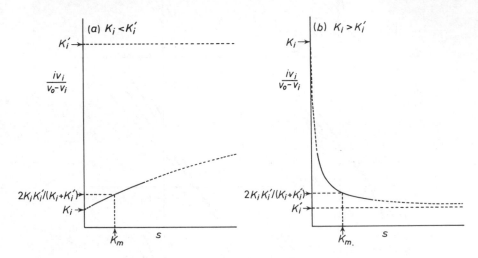

Figure 4.3 Determination of both K_i *and* K_i' *from a single plot of* $iv_i/(v_0 - v_i)$ *against s, for (a)* $K_i < K_i'$ *(predominantly competitive inhibition) and (b)* $K_i > K_i'$ *(predominantly uncompetitive inhibition): In each case, the plot is shown as a broken line except in the range* $s = 0.2K_m$ *to* $s = 2K_m$*. This is done in order to emphasize that unrealistically long extrapolations are required for both inhibition constants to be estimated from this type of plot, unless they are approximately equal*

the intercept and the asymptote accurately, but if K_m is known, one can readily calculate the less accurately defined inhibition constant from the fact that $iv_i/(v_0 - v_i)$ is equal to $2/(1/K_i + 1/K_i')$, i.e. the harmonic mean of K_i and K_i', when $s = K_m$. Alternatively, one can plot $iv_i/(v_0 - v_i)$ against $1/s$ instead of s; in this case, the intercept on the ordinate is K_i'.

It is noteworthy that the right-hand side of equation 4.6 does not contain the inhibitor concentration. Hence velocities can be measured at a haphazard collection of i and s values, but the points should still lie on a single line in a plot of $iv_i/(v_0 - v_i)$ against s. Thus the plot is appropriate for data that could not be plotted in any other way. This may not seem to be a great advantage, but it means that one can explore a much wider range of i and s values than would be possible with a limited number of measurements if one used any of the other plots. This may be useful when it is experimentally difficult to carry out more than a few rate determinations.

All of the plots described in this section are useful, but they are no substitute for computation if accurate values of the inhibition constants are required. Statistical problems arise whenever kinetic parameters are estimated graphically, and may be just as serious in the case of inhibition plots as they are in the simpler cases that have been thoroughly investigated.

4.6 Intuitive approach to linear inhibition

For any mechanism, the most reliable method of determining which of the kinetic parameters is likely to be affected by an inhibitor is to derive the appropriate rate equation and examine it critically. Nonetheless, a more intuitive approach may also be helpful as a mnemonic. In order to introduce this, it is useful to examine the intuitive meanings of K_m, V and V/K_m. K_m was originally used as a measure of the binding of substrate to the enzyme and, although this interpretation of K_m must be used very cautiously, it is adequate for the present purpose. An inhibitor molecule that binds to the same site on the enzyme as the substrate must plainly reduce the capacity of the enzyme to bind substrate, and so must increase K_m^{app}. Provided that it has no other effect, it cannot alter the reactivity of any ES molecules that are formed, and thus cannot affect V^{app}. The term competitive is obviously appropriate for this type of inhibitor. In a sense, it is unfortunate that the term is so appropriate because, if it were not so, it is doubtful whether the meaningless terms non-competitive and uncompetitive would ever have become current. These two terms are best regarded simply as labels, with no connection between form and meaning.

V is a measure of the rate that would result if the enzyme existed wholly as ES complex. Any inhibitor that interferes with the breakdown of ES to products, whether by binding to one of the intermediates to form a dead-end complex or by reversing one of the steps by the law of mass action, must therefore reduce V^{app}. These effects will normally reduce the relative amounts of free enzyme and ES complex, however, and will therefore also alter K_m^{app}. Thus any compound that interferes with the breakdown of ES to products may be expected to be a mixed inhibitor.

V/K_m is often regarded simply as a derived quantity, the ratio of V and K_m, but it also has a much more fundamental meaning: it is the pseudo-first-order rate constant for the reaction

$$E + S \rightarrow E + P$$

The Michaelis–Menten mechanism simplifies to this mechanism if the concentration of the ES complex is insignificant, i.e. at very low substrate concentrations. No saturation behaviour can be detected under such conditions because the small number of ES complexes that are formed exist for such a short time that they do not reduce the free enzyme concentration significantly, and so do not alter the probability of collision between free enzyme and substrate significantly. It is for this reason that at low substrate concentrations ($s < 0.1K_m$), a plot of v against s approximates to a straight line of slope V/K_m. The importance of this approximation in the context of inhibition is that the characteristic property of an uncompetitive inhibitor is that it has

61

no effect on V^{app}/K_m^{app}. It is clear from the above discussion that uncompetitive inhibitors characteristically have negligible effects at very low concentrations of ES. When this is realized, it is not difficult to understand why uncompetitive inhibition arises in the particular mechanisms that it does. As has been seen in the previous section, there are two of these mechanisms: in the first, the inhibitor binds exclusively to the ES complex or another intermediate, but not to the free enzyme. However, at very low substrate concentrations, the free enzyme is the only enzyme species present at any significant concentration, and so an inhibitor that binds only to other forms cannot have any effect. The second (and more important) situation where uncompetitive inhibition occurs is in cases of product inhibition where the reversible substrate-binding and product-release steps (shown as double-headed arrows) are isolated from one another on both sides by irreversible steps (shown as single-headed arrows), e.g.

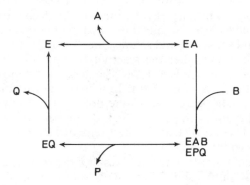

In this mechanism, P can inhibit only by binding to EQ. This is possible at high concentrations of A when one of the product-release steps is rate determining, but it is impossible at low concentrations of A, when $E + A \rightarrow EA$ becomes the rate-determining step, and E is the only significant enzyme species. Thus P is an uncompetitive inhibitor in this mechanism, as it inhibits only at high concentrations of A. Individual steps in a mechanism can become irreversible either because a substrate is present at a saturating concentration (as B in this example) or because a product is present at a zero concentration (as Q at zero time in this example).

4.7 Hyperbolic inhibition and activation

Most of the enzyme inhibitors that have been studied have been interpreted as linear inhibitors, i.e. one of the types considered in the earlier part of this chapter. Nonetheless, it is likely that there are many more exceptions than have been recognized, because the Botts–Morales scheme is very plausible and it does not, in general, predict linear inhibition except under rather restrictive conditions. When considering the complete Botts–Morales scheme, it is convenient to treat inhibition and activation together, because the difference between them is quantitative rather than qualitative and the same algebra applies to both. The symbol X will be used to represent any modifier, whether it be an inhibitor or an activator. Then, if all of the binding steps in the reaction are treated as equilibria, with dissociation constants as indicated in the following scheme:

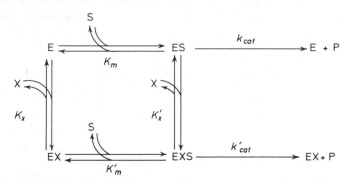

the rate equation is

$$v = \frac{(k_{cat} + k'_{cat}x/K'_x)e_0 s}{K_m(1 + x/K_x) + s(1 + x/K'_x)} \tag{4.7}$$

Actually, this equation applies in practice even if the binding steps are not at equilibrium in the steady state, because the deviations from it that occur in this case are usually too small to be detected. Hence K_m, K_x and K'_x cannot be interpreted as equilibrium constants, even though they were assumed to be so in deriving the equation.

Equation 4.7 is of the form of the Michaelis–Menten equation, with

$$V^{app} = \frac{(k_{cat} + k'_{cat}x/K'_x)e_0}{1 + x/K'_x} \tag{4.8}$$

$$K_m^{app} = \frac{K_m(1 + x/K_x)}{1 + x/K'_x} \tag{4.9}$$

$$V^{app}/K_m^{app} = \frac{(k_{cat} + k'_{cat}x/K'_x)e_0}{K_m(1 + x/K_x)} \tag{4.10}$$

V^{app} increases with increasing x if $k'_{cat} > k_{cat}$. So at high s, when $v \to V^{app}$, X is an activator if $k'_{cat} > k_{cat}$ but an inhibitor if $k'_{cat} < k_{cat}$. At low s, $v \to V^{app}s/K_m^{app}$, and so X affects the velocity according to equation 4.10: in this case, V^{app}/K_m^{app} increases with x if $k'_{cat}/K'_x > k_{cat}/K_x$. It is therefore possible for a given modifier to be an inhibitor at low s and an activator at high s, or *vice versa*. Hence the distinction between activators and inhibitors becomes blurred when the full Botts–Morales scheme applies.

Equations 4.8–4.10 are all of the same general form and in each case, if the left-hand side of the equation is plotted against either x or $1/x$, the result is a rectangular hyperbola that does not pass through the origin. No amount of algebraic manipulation can convert these plots into straight lines, because each contains three independent constants. However, the curves can be analysed approximately by the method indicated previously in *Figure 4.3* for a curve of the same form, although in general computation is more reliable (*see* Chapter 10). Because of the shapes of these curves, modifiers that require the full Botts–Morales scheme are often called *hyperbolic* activators or inhibitors. Some workers use the term *partial* inhibition, to indicate that enzyme activity is not abolished totally at a saturating concentration of inhibitor.

Certain special cases are noteworthy. If $k'_{cat} = 0$, the equations reduce to those for linear mixed inhibition (Section 4.4). If $k'_{cat} = k_{cat}$, we have hyperbolic competitive activation or inhibition. In this case, the term competitive is not really appropriate, as S and X can bind simultaneously and so hardly compete with one another. The term comes from the fact that equation 4.8 reduces to $V^{app} = k_{cat}e_0 = V$, just as in linear competitive inhibition. The two cases can be distinguished by the fact that in hyperbolic competitive inhibition V^{app}/K_m^{app} does not approach zero when $x \to \infty$, and a plot of K_m^{app}/V^{app} against x is a hyperbola and not a straight line.

In order to diagnose hyperbolic inhibition clearly, it is necessary to measure the inhibition behaviour at several values of i spread over a wide range. Failure to do this is probably the main reason why hyperbolic inhibition has been reported so rarely. Only in two (admittedly common) cases can inhibitors be reasonably expected to be linear inhibitors, viz. inhibition by products of the reaction (*see* Chapter 5) and by close substrate analogues, i.e. compounds that are so similar to the substrate that competition for the same binding site is inevitable. Even close substrate analogues may be hyperbolic inhibitors, e.g. the inhibition of alcohol dehydrogenase by methanol (Wratten and Cleland, 1965). In this case, the binding site is presumably capacious enough to accommodate ethanol and methanol at the same time. For all other types of inhibitor, whether physiologically significant or not, hyperbolic inhibition should be regarded as the norm and linear inhibition a deviation from it (this attitude has not been common in the past, however).

Enzymes that occupy key positions in the control of metabolism are often found to be inhibited or activated by compounds that bear no structural similarity to the substrates or products of the reaction. In many cases, there is good evidence that the binding sites for substrate and modifier are separate. This phenomenon is often called *allosteric* inhibition or activation, and bears an obvious similarity to hyperbolic inhibition or activation. However, allosteric enzymes often display other complex properties also, and are discussed in Chapter 7. In the context of metabolic control, the term *effector* is often used instead of modifier, in order to indicate more clearly that the effects are assumed to have a physiological significance.

4.8 Non-productive binding

Much of the information that exists about the general properties of enzymes has been obtained from the study of a small group of enzymes, the extracellular hydrolytic enzymes, including pepsin, lysozyme, ribonuclease and, most notably, chymotrypsin. These enzymes share various properties that make them eminently suitable for detailed study: they are abundant, easily crystallized,* stable, monomeric and can be treated as single-substrate enzymes, as the second substrate is water in each case. However, they also share the disadvantage, which must always be remembered when interpreting results, that they are all usually studied with simple artificial substrates that are much less bulky than their ill-defined and polymeric natural substrates. However, an enzyme that is capable of binding a polymer is likely to be able to bind a small molecule in many ways. Thus, instead of a single enzyme–substrate complex that breaks down to products, there may be in addition numerous 'non-productive complexes' that do not break down. This is illustrated in the following scheme:

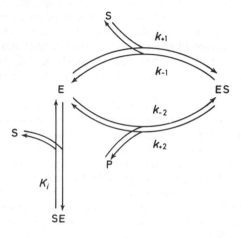

where SE represents all non-productive complexes. This scheme is the same as that for linear competitive inhibition (Section 4.2) with the inhibitor replaced with substrate, and the rate equation (cf. equation 4.1) is

$$v = \frac{k_{+2}e_0 s}{\left(\dfrac{k_{-1}+k_{+2}}{k_{+1}}\right)\left(1 + \dfrac{s}{K_i}\right) + s} \qquad (4.11)$$

If the *expected* values of V and K_m are defined as the values they would have if no non-productive complexes were formed, i.e. $V^{\text{exp}} = k_{+2}e_0$, $K_m^{\text{exp}} = (k_{-1}+k_{+2})/k_{+1}$ (cf. 'pH-corrected' constants, Section 6.3), then equation 4.12 can be rearranged to give

$$v = \frac{Vs}{K_m + s}$$

---

* But crystallinity is no longer regarded as a certain guarantee of purity for enzymes; for example, both trypsin and urease have been available for many years in the crystalline, but far from pure, state.

where

$$V = \frac{V^{\exp}}{1 + K_m^{\exp}/K_i}$$

$$K_m = \frac{K_m^{\exp}}{1 + K_m^{\exp}/K_i}$$

$$V/K_m = V^{\exp}/K_m^{\exp}$$

Thus the Michaelis–Menten equation is obeyed exactly for this mechanism and so the observed kinetics do not indicate whether non-productive binding is significant or not. Unfortunately, it is often the expected values that are of interest in an experiment, because they refer to the main productive catalytic pathway. Hence the measured values of V and K_m may be less, by an unknown and unmeasurable amount, than the quantities of interest. Only V/K_m gives a correct measure of the catalytic properties of the enzyme.

For highly specific enzymes, plausibility arguments can be used to justify the exclusion of non-productive binding from consideration, but for unspecific enzymes, such as chymotrypsin, comparison of the results for different substrates can sometimes provide evidence of the phenomenon. For example, Ingles and Knowles (1967) measured the rates of hydrolysis of a series of acylchymotrypsins. They found, after allowing for differences in the inherent reactivity of the acyl groups, that for derivatives of L-amino acids, such as acetyl-L-tryptophanylchymotrypsin, the rate was fastest with large hydrophobic groups, but for the corresponding derivatives of D-amino acids the opposite result was observed. The simplest interpretation is in terms of non-productive binding: for acyl groups with the correct L configuration, the large hydrophobic side-chains permit tight and rigid binding in the correct mode, largely ruling out non-productive complexes; but for acyl groups with the D configuration, the same side-chains favour tight and rigid binding in non-productive modes.

Non-productive binding is not usually considered in the context of inhibition, indeed, it is usually not considered at all, but it is plainly a special type of competitive inhibition and it is important to be aware of it when interpreting results for several substrates of an unspecific enzyme. The term substrate inhibition is usually reserved for the uncompetitive analogue of non-productive binding, which is considered in the next section.

4.9 Substrate inhibition

For some enzymes it is possible for a second substrate molecule to bind to the enzyme–substrate complex, ES, to produce an inactive complex, SES, as shown at the top of p. 67. This scheme is analogous to that for uncompetitive inhibition (Section 4.5) and gives the following equation for the initial rate:

$$v = \frac{k_{+2}e_0 s}{\left(\dfrac{k_{-1}+k_{+2}}{k_{+1}}\right) + s(1 + s/K_{si})} = \frac{Vs}{K_m + s + s^2/K_{si}}$$

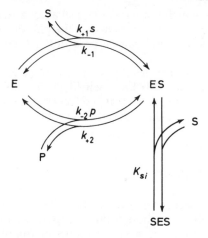

where V and K_m are defined in the usual way as $k_{+2}e_0$ and $(k_{-1}+k_{+2})/k_{+1}$, respectively. This equation is not of the form of the Michaelis–Menten equation, by virtue of the term in s^2. This term becomes significant only at high substrate concentrations. Hence the velocity approaches the Michaelis–Menten value when s is small, but approaches zero instead of V when s is large. By differentiating with respect to s and putting dv/ds to zero, it can readily be shown that the maximum velocity (not equal to V) occurs when $s^2 = K_m K_{si}$. The curve of v against s is illustrated in *Figure 4.4* together with

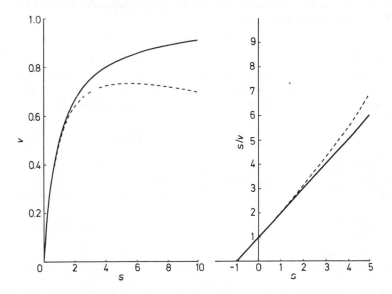

Figure 4.4 Effect of substrate inhibition on plots of v *against* s *and of* s/v *against* s*: In both plots, the solid lines are calculated with* $K_m = 1$, $V = 1$, *with no inhibition, and the broken lines are calculated with the same values of* K_m *and* V, *and* $K_{si} = 30$

a plot of s/v against s, which is a parabola instead of a straight line. Provided that K_{si} is much larger than K_m (as is usual), the plot of s/v against s is almost straight at low values of s, and can be used in the usual way to estimate V and K_m.

Substrate inhibition is not usually a significant phenomenon if substrate concentrations are kept at or below their likely physiological values, but it can become important at high substrate concentrations and provides a useful diagnostic tool for distinguishing between possible pathways, as is discussed in Section 5.6.

4.10 Inhibitors of high affinity

At the beginning of this chapter, a distinction was drawn between reversible and irreversible inhibitors. Although this is a useful distinction to make, it is theoretically objectionable because it implies a qualitative and absolute difference when actually many 'irreversible' inhibitors are simply reversible inhibitors with a very high affinity for the enzyme. Straus and Goldstein (1943) showed that it is possible to treat inhibitors of high affinity without introducing any arbitrary qualitative assumptions. It is simplest to consider first the case of an inhibitor binding to an enzyme in the absence of substrate:

$$\begin{array}{ccc} E + I & \rightleftarrows & EI \\ e_0-y \quad i_0-y & K_i & y \end{array}$$

Note that the free inhibitor concentration is not assumed to be the same as the total inhibitor concentration, i_0. At equilibrium, the concentration, y, of the complex EI is given by

$$y = (e_0 - y)(i_0 - y)/K_i \qquad (4.12)$$

This equation can be rearranged to give a quadratic equation for y, but it is more useful to consider the fraction, α, of total enzyme in unbound form, i.e. $\alpha = (e_0 - y)/e_0$, as α is more likely than y to be directly measurable. Eliminating y from equation 4.12, and rearranging, we obtain

$$i_0 = (1-\alpha)e_0 + \left(\frac{1}{\alpha} - 1\right)K_i \qquad (4.13)$$

This equation now provides the value of i_0 necessary to give any value of α. Although it can be rearranged so as to show α in terms of i_0, the result is much less manageable than the form shown. Equation 4.13 simplifies in two limiting cases:

if $e_0 \ll K_i$,

$$i_0 = \left(\frac{1}{\alpha} - 1\right)K_i \text{ or } \alpha = \frac{1}{1 + i_0/K_i} \qquad (4.14)$$

if $e_0 \gg K_i$,

$$i_0 = (1-\alpha)e_0 \text{ or } \alpha = 1 - i_0/e_0 \qquad (4.15)$$

Thus the shape of the binding curve is a function of the enzyme concentration. Straus and Goldstein defined three 'zones' of inhibition behaviour, namely Zone A for the case when equation 4.14 applies, Zone C for the case when equation 4.15 applies and Zone B for the intermediate case when the full equation 4.13 is required. A range of binding curves is shown in *Figure 4.5*. For

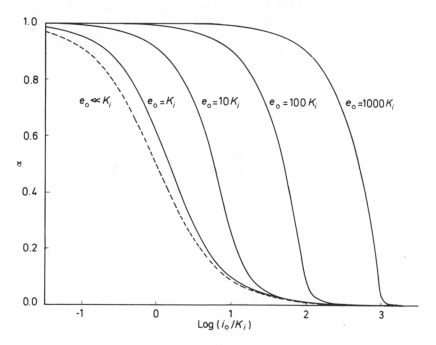

$$e_o \ll K_i \qquad e_o = K_i \qquad e_o = 10 K_i \qquad e_o = 100 K_i \qquad e_o = 1000 K_i$$

Figure 4.5 Fraction of enzyme in unbound form, α, expressed as a function of total inhibitor concentration, i_0, for the system $E + I \rightleftarrows EI$, with dissociation constant K_i, and various values of the total enzyme concentration, e_0, as indicated

small values of e_0 (less than about $0.1K_i$), the shape of the curve is independent of K_i, as expected from equation 4.14. In steady-state kinetic experiments, enzyme concentrations are generally very low, typically in the range 10^{-10}–10^{-7} M. Consequently, only inhibitors of very high affinity deviate significantly from equation 4.14 in such experiments. However, a different situation probably applies in living cells: Srere (1968), for example, has found the cellular concentrations of several important enzymes to be in the range 10^{-7}–10^{-4} M. Zone A behaviour may therefore be much less common in metabolism than it is in the test-tube, and many substrates and inhibitors may exist to a significant extent as enzyme-bound species. High enzyme concentrations are also usual in equilibrium binding studies and in transient-state kinetic experiments; in both cases, it is likely to be invalid to assume that free and total inhibitor or substrate concentrations are equal.

Equation 4.13 was derived for the equilibrium case, with no substrate present. If substrate is added, the behaviour becomes much more complicated, because in all types of inhibition except pure non-competitive in-

69

hibition the addition of substrate perturbs the binding of inhibitor. If the binding of inhibitor is very rapid, then there is no alternative to a full solution of the complicated equations that describe competitive inhibition in Zone B. Fortunately, in at least some cases involving inhibitors with high affinity, the release of inhibitor from the EI complex is slow enough for it to be possible to ignore perturbation of the equilibrium by addition of substrate. Thus, for example, Myers (1952) was able to treat the inhibition of pseudo-cholines-terase by 'Nu 683,' a potent competitive inhibitor of the enzyme, by means of equation 4.13. Effectively, therefore, the slowness of the reaction between the enzyme and inhibitor made the inhibition appear to be non-competitive. By considering only the case of 50% inhibition, i.e. $\alpha = 0.5$, Myers converted equation 4.13 into the very simple expression

$$i_{0.5} = K_i + 0.5e_0$$

where $i_{0.5}$ is the value of i_0 necessary for $\alpha = 0.5$. Of course, this approach is valid only if the rate can be measured immediately after addition of substrate to the enzyme, and before the inhibitor is significantly displaced.

5

Reaction Pathways

5.1 Introduction

Much of the earlier part of this book has been concerned with reactions of a single substrate and a single product. Actually, such reactions are rather rare in biochemistry, being confined to a few isomerizations, such as the interconversion of glucose 1-phosphate and glucose 6-phosphate, catalysed by phosphoglucomutase. In spite of this, the development of enzyme kinetics was greatly simplified by two facts: firstly, the many hydrolytic enzymes can normally be treated as single-substrate enzymes, because the second substrate, water, is always present in such large excess that its concentration can be treated as a constant; secondly, most enzymes behave much like single-substrate enzymes if only one substrate concentration is varied. This will be clear from the rate equations to be introduced in this chapter, but the proviso exists in this case that K_m for a single substrate has a physical meaning only if the constant conditions are both well defined and constant.

The mechanisms of single-substrate reactions are not entirely trivial, because most realistic models of catalysis of isomerization require that the free enzyme be released in a different form from that which bound the substrate. Instead of the simple Michaelis–Menten mechanism, therefore, the mechanism is likely to be

$$E + S \rightleftarrows ES \rightleftarrows E' + P$$

The final isomerization, $E' \rightarrow E$, is in principle detectable by product-inhibition studies; but complications arise from the fact that it is impossible to prevent the back reaction from occurring when product inhibition is tested in a one-product reaction. This is discussed further in Section 5.10.

There are three principal kinetic methods for elucidating reaction pathways: measurement of initial rates in the absence of product; testing the nature of product inhibition; and tracer studies with radioactively labelled substrates. These methods are discussed in this chapter, using a general two-substrate, two-product reaction as an example:

$$A + B \rightleftarrows P + Q$$

This equation represents by far the commonest type of reaction in bio-chemistry, as it describes about 60% of all known enzyme-catalysed reactions. More complex reactions exist, with as many as four or more substrates, but these reactions can be studied by a simple extension of the principles developed for the study of two-substrate, two-product reactions.

5.2 Survey of two-substrate, two-product reaction mechanisms

Almost all two-substrate, two-product reactions are formally *group-transfer* reactions, i.e. reactions in which a group, G, is transferred from one radical, X, to another, Y:

$$GX + Y \rightleftarrows X + GY$$

Wong and Hanes (1962) suggested that most reasonable possibilities for this transfer would be encompassed by the following scheme:

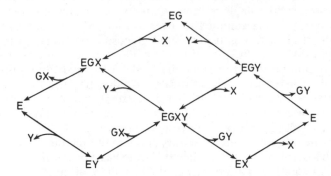

It would not be expected that all of the steps shown would occur with any one enzyme, and indeed it is fortunate that very few enzymes follow such a complex scheme: the King–Altman analysis requires 96 patterns and gives a steady-state rate equation of extreme complexity, containing, for example, terms in $[Y]^3$. Some biotin-dependent enzymes, such as methylmalonyl-CoA carboxyltransferase (Northrop, 1969) and pyruvate carboxylase (Barden *et al.*, 1972), seem able to undergo most or all of the reactions in the Wong–Hanes scheme; but a single pathway, via EG, is strongly preferred, and many of the reactions can be treated as equilibria (cf. Section 3.6), so that experimentally the kinetic behaviour of these enzymes is much simpler than a rigorous equation would suggest.

Most enzymes obey mechanisms that are much simpler than the Wong–Hanes scheme. The main division is between mechanisms that proceed through a *ternary complex*, EGXY, so called because it contains the enzyme and both substrates in a single species, and those that proceed through a

72

substituted enzyme, EG. Early workers, such as Woolf (1929, 1931) and Haldane (1930), assumed that the reaction would proceed through a ternary complex, and that this could be formed by way of either of the two binary complexes, EGX and EY. In other words, the substrates could bind to the enzyme in *random order*, as illustrated in *Figure 5.1*. The rigorous steady-state equation for this mechanism is complex, and includes terms in $[GX]^2$ and $[Y]^2$. The contribution of such terms to the rate is very slight, however, and Gulbinsky and Cleland (1968) have shown by computer simulation that unless very implausible values are assumed for the rate constants the experimental rate equation is of the same form as one derived on the assumption that all steps except the interconversion of $EXG \cdot Y$ and $EX \cdot GY$ are at equilibrium. If this assumption is made, no square terms appear in the rate equation, and for simplicity we shall use rate equations (for random-order mechanisms only) derived with the rapid-equilibrium assumption. However, it must be emphasized that the fact that such equations are obeyed experimentally does *not* imply that the equilibrium assumption is correct, any more than the fact that most enzymes obey the Michaelis–Menten equation implies that the Michaelis–Menten assumption of equilibrium binding is usually correct. The step $EXG \cdot Y \rightleftarrows EX \cdot GY$ cannot be detected by steady-state measurements (cf. Section 3.7), but it is logical to include it in the random-order mechanism as it is formally treated as rate-determining in deriving the rate equation.

The non-productive complex, EXY, is not a necessary feature of the random-order mechanism, but it can normally be expected to occur, because if both EY and EX are significant intermediates there is no reason to exclude EXY. Another non-productive complex (not included in *Figure 5.1*) can occur if the transferred group G is not too bulky: $EXG \cdot GY$ can result from the binding of GY to EGX or of GX to EGY. This is less likely than the formation of EXY, however.

It is now generally recognized that many enzymes cannot be regarded as rigid templates, as suggested by *Figure 5.1*. Instead, it is likely that the conformations of both enzyme and substrate are altered upon binding, in accordance with the 'induced-fit' hypothesis (Koshland, 1958, 1959*ab*; *see also* Section 7.6). Consequently, it may well happen that no binding site exists on the enzyme for one of the two substrates until the other is bound. In such cases, there is a *compulsory order* of binding, as illustrated in *Figure 5.2*. (Actually this is reconcilable with a rigid-template model if the second substrate interacts strongly with the first substrate as well as with the enzyme. However, this would usually lead to a random order of binding with one pathway strongly favoured over the other.) If both substrates and products are considered, four different orders are possible, but the induced-fit explanation of compulsory-order mechanisms leads us to expect that the reverse reaction should be structurally analogous to the forward reaction, so that the second product ought to be the structural analogue of the first substrate. Thus only two of the four possibilities are very likely. This is in accordance with observation, and, for example, in NAD-dependent dehydrogenase reactions, the coenzymes are usually found to be first substrate and second product.

73

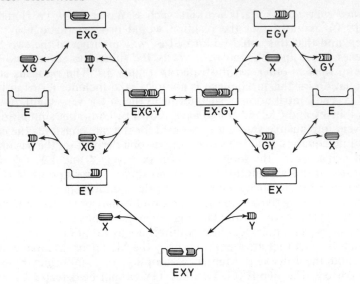

Figure 5.1 Ternary-complex mechanism for a two-substrate, two-product reaction, assuming that the substrates bind to and the products are released from the enzyme in random order: The non-productive complex EXY is likely to be kinetically significant only at high concentrations of both X and Y, and is often ignored in simple treatments

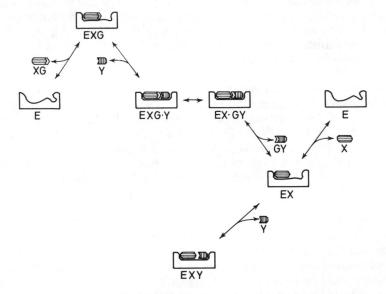

Figure 5.2 Ternary-complex mechanism for a two-substrate, two-product reaction, assuming that the substrates bind to and the products are released from the enzyme in a compulsory order: This is presumed to arise because the binding site for the second substrate becomes recognizable only after an appropriate change in conformation has been induced by the binding of the first substrate. The non-productive complex EXY is likely to be kinetically significant only at high concentrations of both X and Y, and is often ignored in simple treatments

74

For the same reason as in the random-order case, the non-productive complex EXY is commonly observed in compulsory-order mechanisms, as a result of binding of X to EY, or of Y to EX.

At an early stage in the development of multiple-substrate kinetics, Doudoroff, Barker and Hassid (1947) showed by isotope-exchange studies that the reaction catalysed by sucrose glucosyltransferase proceeded through a substituted-enzyme intermediate rather than a ternary complex. Since then, studies with numerous enzymes, including α-chymotrypsin, trans-aminases and flavoenzymes, have shown that the substituted enzyme is a very important and common alternative to the ternary complex. The substituted-enzyme mechanism is included in the Wong–Hanes scheme, and is shown schematically in *Figure 5.3*. In the normal ('classical') form of this mechanism,

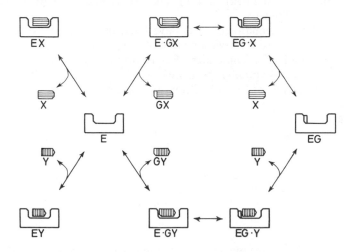

Figure 5.3 Substituted-enzyme mechanism for a two-substrate, two-product reaction: The sites for X and Y are assumed to coincide or overlap, in contrast to the case for ternary-complex mechanisms (Figures 5.1 and 5.2). The non-productive complexes EX and EY are likely to be kinetically significant only at high concentrations of X and Y, respectively, and are often ignored in simple treatments

the formation of ternary complexes is prevented by the fact that the binding sites for X and Y are either the same or overlapping. For the transaminases, a major group of enzymes that obey this mechanism, all four reactants are structurally similar, so that it is reasonable to expect the binding sites for X and Y to be virtually identical and the second half of the reaction is there-fore very similar to the reverse of the first half, e.g.

glutamate + pyridoxal–enzyme $\rightleftarrows$ intermediate(s)
$$\rightleftarrows \text{α-ketoglutarate} + \text{pyridoxamine–enzyme}$$

oxaloacetate + pyridoxamine–enzyme $\rightleftarrows$ intermediate(s)
$$\rightleftarrows \text{aspartate} + \text{pyridoxal–enzyme}$$

In this mechanism, it is usually possible for substrates to bind to the 'wrong' form of the enzyme, resulting in substrate inhibition at high concentrations

(Section 5.6). This is almost always true of E, X and Y, but less often of EG, GX and GY because of steric interference between two G groups.

The substituted-enzyme mechanism is also a compulsory-order mechanism, but this is less important than with ternary-complex mechanisms because there is only one possible order, and no random-order alternative: although E can often bind X or Y, there is no way for the resulting complexes to break down to give GX or GY.

As mentioned earlier, certain reactions catalysed by biotin-containing enzymes proceed predominantly by a substituted-enzyme mechanism, but these are atypical in that the sites for X and Y are independent, and ternary complexes occur as alternative intermediates. In the rest of this chapter (and in the literature), the normal form of the substituted-enzyme mechanism, with overlapping or identical binding sites for X and Y, is to be assumed unless otherwise stated.

With any of the mechanisms that have been discussed, it is possible to assume as a special case that two or more of the steps are concerted, i.e. that they can be treated as a single step. For example, the Doudoroff–Barker–Hassid mechanism, originally proposed (1947) for sucrose glucosyltransferase, is a special case of the substituted-enzyme mechanism:

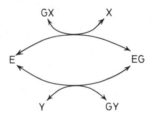

and the Theorell–Chance mechanism, originally proposed (1951) for alcohol dehydrogenase, is a similar special case of the compulsory-order ternary-complex mechanism:

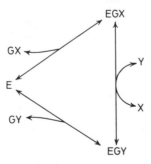

Although no species EGXY occurs as an intermediate in this mechanism, it must occur as a transition state, and this mechanism is no less a ternary-complex mechanism than the usual types.

The concerted mechanisms are mainly of historic interest, because it is now usually possible to detect the existence of the omitted intermediates, but they also provide a conveniently simple formulation for examining the chemistry of the two types of mechanism. In the ternary-complex mechanism, a direct reaction between the two substrates takes place on the enzyme surface, as the attacking group Y displaces the leaving group X:

Although the enzyme appears to have no direct role in this mechanism, it undoubtedly fulfils two important functions: it constrains the two substrates to adopt suitable conformations and locations for reaction, and it polarizes the electrons in the substrates so as to make them more reactive. Koshland (1954) called this a *single-displacement* reaction. He pointed out its close analogy with the well-known S_N2 (bimolecular nucleophilic substitution) reactions in organic chemistry and suggested that single-displacement reactions would normally be accompanied by inversion of the configuration at the substituted atom.

Koshland referred to the substituted-enzyme mechanism as a *double-displacement* reaction, as an initial displacement of X by an attacking group B on the enzyme is followed by a displacement of B by the second substrate Y:

This mechanism is analogous to reactions that involve 'neighbouring-group effects' in organic chemistry. The net effect of two inversions of configuration is to regenerate the original configuration at the substituted atom.

Retention of configuration is also possible in a single-displacement reaction, if the attacking group approaches on the same side as the leaving group ('frontside attack'). This is the S_Ni (internal nucleophilic substitution) reaction and is extremely rare in organic chemistry, being largely confined to reactions that involve thionyl chloride or phosgene, but may be more

77

common in enzyme-catalysed reactions, given the unusual conditions that presumably exist on the enzyme surface.

The stereochemistry of enzyme-catalysed reactions provides a useful tool for investigating reaction mechanisms, and is complementary to the kinetic methods that are discussed later in this chapter. However, kinetic and stereochemical tests need not give identical results: not only can single-displacement reactions result in retention of configuration, in the event of frontside attack, but also double-displacement reactions can obey ternary-complex kinetics, if the first leaving group X remains bound to the enzyme until after the second displacement. A reasonable mechanism in which this would happen would be one in which X remained attached to an acidic group A^+ as long as the basic group B^- was unavailable to form an ionic bond with A^+:

The system of Wong and Hanes for representing group-transfer reactions with the symbols X, Y and G is very useful in the qualitative discussion of mechanisms, as in this section, because it provides a very clear picture of each detailed mechanism. However, for a quantitative description of kinetics it is rather less satisfactory, because it does not distinguish in an obvious way between substrates and products, and it does not lend itself to a short-hand representation of reactant concentrations. For the remainder of this chapter, therefore, we shall return to the use of single letters, A, B ..., P, Q ... to represent reactants. In the compulsory-order ternary-complex mechanism A and Q are defined as the reactants that bind to the free enzyme, in the random-order mechanism the labelling of reactants is arbitrary, and in the substituted-enzyme mechanism A and Q bind to E, and B and P to E', although it is arbitrary which way round E and E' are defined. These rules can be extended in an obvious way to mechanisms that involve more than two substrates or products.

5.3 Nomenclature and schematic representation of mechanisms

Cleland (1963) has proposed a system for representing mechanisms schematically. The various forms of the enzyme are written below a line, and arrows are drawn to show the addition of substrates and release of products. Thus the compulsory-order ternary-complex mechanism is shown as

Random-order steps are accommodated by a branched line, as in the random-order ternary-complex mechanism:

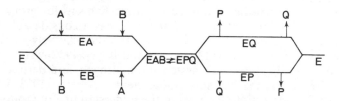

The forward reaction is found by reading from left to right and following the arrows, and the reverse reaction by reading from right to left and reversing the arrows. This system provides a compact method of representing mechanisms, and is particularly convenient for simple compulsory-order mechanisms. It is less convenient for mechanisms with random-order steps and inhibitory side reactions, and does not readily lend itself to the inclusion of rate constants.

Cleland (1963) has also proposed a general nomenclature for enzyme mechanisms. Firstly, all two-substrate, two-product reactions are called *bi bi* reactions. Of these, the random-order ternary-complex mechanism is called simply *random bi bi*; the compulsory-order ternary-complex mechanism is called *ordered bi bi*, although this does not distinguish it (except by convention) from the substituted-enzyme mechanism, which is called *ping pong bi bi*. All mechanisms that require binding of every substrate before any product can be released (i.e. ternary-complex mechanisms in the two-substrate, two-product case) are called *sequential* mechanisms. Conversely, mechanisms in which some products are released before every substrate has bound are called *ping pong* mechanisms. Mechanisms that involve isomerization of the free enzyme are called *iso* mechanisms. The terms *uni, bi, ter, quad* are used in ambiguous cases in order to define the numbers of substrate additions and product dissociations.

The Cleland system can, in principle, be applied to mechanisms of great complexity, although in practice a description or a diagram is clearer. For example, a definition such as *iso bi bi uni uni ping pong* is unlikely to be understood as easily as 'a three-substrate, three-product mechanism in which the binding of the first two substrates in compulsory order is followed by the release of two products, then the binding of the third substrate, release of the third product, and finally isomerization of the enzyme to its original form,' and a diagram is clearer than either:

79

5.4 Rate equations

Steady-state kinetic measurements have proved to be of enormous value in distinguishing between the various reaction mechanisms for group-transfer reactions. The development of these methods was a considerable task, on account of the large number of possibilities and the relatively small kinetic differences between them. Segal, Kachmar and Boyer (1952) were among the first to recognize the need for a systematic approach, and derived the rate equations for several mechanisms. Subsequently, Alberty (1953, 1958) and Dalziel (1957) made major advances in the understanding of group-transfer reactions, and introduced most of the methods described in this chapter.

As all steady-state methods for distinguishing between mechanisms depend on differences between the complete rate equations, it is appropriate to give a brief account of these equations before discussing methods. The equation for the compulsory-order ternary-complex mechanism was derived in Section 3.3 as an illustration of the King–Altman method, giving

$$v = \frac{e_0(c_1 ab - c_2 pq)}{c_3 + c_4 a + c_5 b + c_6 p + c_7 q + c_8 ab + c_9 ap + c_{10} bq + c_{11} pq + c_{12} abp + c_{13} bpq} \qquad (5.1)$$

This equation contains thirteen coefficients, but these were defined in terms of only eight rate constants, and so there must be relationships between the coefficients that are not explicit in the equation. Moreover, the coefficients are without obvious meaning. Numerous systems have been used for re-writing rate equations in more meaningful terms (*see*, for example, Alberty, 1953; Dalziel, 1957; Bloomfield, Peller and Alberty, 1962; Cleland, 1963; Mahler and Cordes, 1966). Of these, the simplest to understand and use is probably that of Cleland, modified slightly in this book to accord with the recommendations of the Enzyme Commission of the International Union of Biochemistry (1961). For any mechanism, maximum velocities in the forward and reverse directions are written as V^f and V^r, respectively, although the superscripts can be omitted in unambiguous cases; in addition, for each reactant, a Michaelis constant, K_m^A, K_m^B, etc., and an 'inhibition' constant, K_i^A, K_i^B, etc., are defined. The meanings of these will become clear in subsequent sections of this chapter, but in general the Michaelis constants correspond to K_m in a one-substrate reaction, and the inhibition constants are related to (but not necessarily equal to) the K_i or K_i' values measured in product-inhibition experiments. Under certain circumstances, the inhibition constants are true substrate-dissociation constants, and some workers therefore write K_s rather than K_i for them.

With this system, equation 5.1 becomes

$$v = \frac{\dfrac{V^f ab}{K_i^A K_m^B} - \dfrac{V^r pq}{K_m^P K_i^Q}}{1 + \dfrac{a}{K_i^A} + \dfrac{K_m^A b}{K_i^A K_m^B} + \dfrac{K_m^Q p}{K_m^P K_i^Q} + \dfrac{q}{K_i^Q} + \dfrac{ab}{K_i^A K_m^B} + \dfrac{K_m^Q ap}{K_i^A K_m^P K_i^Q} + \dfrac{K_m^A bq}{K_i^A K_m^B K_i^Q} + \dfrac{pq}{K_m^P K_i^Q} + \dfrac{abp}{K_i^A K_m^B K_i^P} + \dfrac{bpq}{K_i^B K_m^P K_i^Q}}$$

(5.2)

where the kinetic parameters have the values shown in *Table 5.1*.

Table 5.1 DEFINITIONS OF KINETIC PARAMETERS FOR THE TWO PRINCIPAL COMPULSORY-ORDER MECHANISMS

Parameter	Ternary-complex mechanism	Substituted-enzyme mechanism
	$E \underset{k_{-1}}{\overset{k_{+1}a}{\rightleftharpoons}} EA$ $k_{+4}\Vert k_{-4}q \quad k_{-2}\quad k_{+2}b\Vert$ $EQ \underset{k_{+3}}{\overset{k_{-3}p}{\rightleftharpoons}} \begin{array}{c}EAB\\EPQ\end{array}$	$E \underset{k_{-1}}{\overset{k_{+1}a}{\rightleftharpoons}} \begin{array}{c}EA\\E'P\end{array}$ $k_{+4}\Vert k_{-4}q \quad k_{-2}p\quad k_{+2}\Vert$ $\begin{array}{c}EQ\\E'B\end{array} \underset{k_{+3}b}{\overset{k_{-3}}{\rightleftharpoons}} E'$
V^f	$\dfrac{k_{+3}k_{+4}e_0}{k_{+3}+k_{+4}}$	$\dfrac{k_{+2}k_{+4}e_0}{k_{+2}+k_{+4}}$
V^r	$\dfrac{k_{-1}k_{-2}e_0}{k_{-1}+k_{-2}}$	$\dfrac{k_{-1}k_{-3}e_0}{k_{-1}+k_{-3}}$
K_m^A	$\dfrac{k_{+3}k_{+4}}{k_{+1}(k_{+3}+k_{+4})}$	$\dfrac{(k_{-1}+k_{+2})k_{+4}}{k_{+1}(k_{+2}+k_{+4})}$
K_m^B	$\dfrac{(k_{-2}+k_{+3})k_{+4}}{k_{+2}(k_{+3}+k_{+4})}$	$\dfrac{k_{+2}(k_{-3}+k_{+4})}{(k_{+2}+k_{+4})k_{+3}}$
K_m^P	$\dfrac{k_{-1}(k_{-2}+k_{+3})}{(k_{-1}+k_{-2})k_{-3}}$	$\dfrac{(k_{-1}+k_{+2})k_{-3}}{(k_{-1}+k_{-3})k_{-2}}$
K_m^Q	$\dfrac{k_{-1}k_{-2}}{(k_{-1}+k_{-2})k_{-4}}$	$\dfrac{k_{-1}(k_{-3}+k_{+4})}{(k_{-1}+k_{-3})k_{-4}}$
K_i^A	k_{-1}/k_{+1}	k_{-1}/k_{+1}
K_i^B	$(k_{-1}+k_{-2})/k_{+2}$	$*k_{-3}/k_{+3}$
K_i^P	$(k_{+3}+k_{+4})/k_{-3}$	k_{+2}/k_{-2}
K_i^Q	k_{+4}/k_{-4}	k_{+4}/k_{-4}

* Although equation 5.4 does not contain K_i^B it can be re-written so that it does by means of the identity $K_i^A K_m^B/K_i^P K_m^Q = K_m^A K_i^B/K_m^P K_i^Q$.

The corresponding equation for the random-order ternary-complex mechanism is

$$v = \frac{\dfrac{V^f ab}{K_i^A K_m^B} - \dfrac{V^r pq}{K_m^P K_i^Q}}{1 + \dfrac{a}{K_i^A} + \dfrac{b}{K_i^B} + \dfrac{p}{K_i^P} + \dfrac{q}{K_i^Q} + \dfrac{ab}{K_i^A K_m^B} + \dfrac{pq}{K_m^P K_i^Q}}$$

(5.3)

81

This equation is derived by assuming that all steps other than the inter-conversion of EAB and PQ are at equilibrium. With this assumption, K_i^A, K_i^B, K_i^P and K_i^Q are the dissociation constants of EA, EB, EP and EQ, respectively; K_m^A and K_m^B are the dissociation constants of EAB for loss of A and B, respectively; and K_m^P and K_m^Q are the dissociation constants of EPQ for loss of P and Q, respectively. (Although K_m^A and K_m^Q are absent from equation 5.3, they can be introduced because in this mechanism $K_m^A K_i^B$ is interchangeable with $K_i^A K_m^B$, and $K_i^P K_m^Q$ with $K_m^P K_i^Q$. These substitutions cannot be made in equation 5.2, and it is partly for this reason that equation 5.2 has a more complex appearance, apart from the additional terms that it contains.) However, as equation 5.3 applies within experimental error whether the equilibrium assumption is correct or not, the Michaelis and inhibition constants should not be interpreted as true dissociation constants.

The equation for the substituted-enzyme mechanism is

$$v = \frac{\dfrac{V^f ab}{K_i^A K_m^B} - \dfrac{V^r pq}{K_i^P K_m^Q}}{\dfrac{a}{K_i^A} + \dfrac{K_m^A b}{K_i^A K_m^B} + \dfrac{p}{K_i^P} + \dfrac{K_m^P q}{K_i^P K_m^Q} + \dfrac{ab}{K_i^A K_m^B} + \dfrac{ap}{K_i^A K_i^P} + \dfrac{K_m^A bq}{K_i^A K_m^B K_i^Q} + \dfrac{pq}{K_i^P K_m^Q}} \quad (5.4)$$

where the kinetic parameters are again defined in *Table 5.1*. In coefficient form, this equation is the same as equation 5.1 without the constant and the terms in abp and bpq, but the relationships between the parameters are different, and equation 5.4 has $K_i^P K_m^Q$ wherever $K_m^P K_i^Q$ might be expected from comparison with equation 5.2.

The rate equations are generally similar for the variants of these mechanisms with concerted steps, but certain terms are missing. For example, in the Theorell–Chance mechanism, with the binding of B and release of P in a single step, no term in the rate equation can contain bp as a product. The rate equation for this mechanism is therefore the same as equation 5.2 without the terms in abp and bpq. An interesting prediction results if this mechanism is compressed still further so that Q is released in the same concerted step:

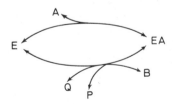

In this case, all of the terms in ab, ap, bq, abp and bpq disappear from the denominator of the rate equation, and the velocity no longer approaches a limiting value if a and b are both increased to large values; instead, it apparently increases to infinity as a and b are made infinite. Catalase (Chance, 1963) appears to obey such a mechanism, and shows no evidence of saturation at high concentrations of both substrates. Similar behaviour is expected for an enzyme that obeys the Doudoroff–Barker–Hassid mechanism, but there do not appear to be experimental examples of this.

5.5 Initial-velocity measurements in absence of products

If no products are included in the reaction mixture, the initial velocity for a reaction following the compulsory-order ternary-complex mechanism is given by the following equation:

$$v = \frac{Vab}{K_i^A K_m^B + K_m^B a + K_m^A b + ab} \tag{5.5}$$

which is obtained from equation 5.2 by omitting terms in product concentrations. The meanings of V, K_i^A, K_m^B and K_m^A become apparent if the equation is examined at extreme values of a and b. If both a and b are very large, the equation reduces to $v = V$, and thus V is the velocity when both substrates are saturating. It corresponds in an obvious way to the maximum velocity V in a single-substrate reaction. If b is very large, equation 5.5 simplifies to

$$v = \frac{Va}{K_m^A + a}$$

i.e. the Michaelis–Menten equation. Hence K_m^A is defined as the limiting Michaelis constant for A when B is saturating. Similarly, K_m^B is the limiting Michaelis constant for B when A is saturating. K_i^A is *not* the same as K_m^A, and its meaning can be seen by considering equation 5.5 when b is very small (but not zero). Then the equation becomes

$$v = \frac{(Vb/K_m^B)a}{K_i^A + a}$$

K_i^A is therefore the limiting value of the Michaelis constant for A when b approaches zero. It is also the true equilibrium dissociation constant of EA, because when b approaches zero the rate of reaction of B with EA must also approach zero; consequently, the binding of A to E can then be maintained at equilibrium, and so the Michaelis–Menten assumption of equilibrium binding is valid in this instance. K_i^B does not appear in equation 5.5, because B does not bind to the free enzyme. Nonetheless, measurement of initial rates in the absence of products does not distinguish A from B, because the form of the equation is unchanged if the substrates are interchanged.

If the concentration of one substrate is varied at constant (but not saturating) concentrations of the other, equation 5.5 still simplifies to the Michaelis–Menten equation; e.g. if a is varied at constant b, we have

$$v = \frac{\left(\dfrac{Vb}{K_m^B + b}\right)a}{\left(\dfrac{K_i^A K_m^B + K_m^A b}{K_m^B + b}\right) + a}$$

but V^{app} and K_m^{app}, the apparent values of V and K_m, depend on the value of b:

$$V^{app} = \frac{Vb}{K_m^B + b} \tag{5.6}$$

$$K_m^{app} = \frac{K_i^A K_m^B + K_m^A b}{K_m^B + b} \tag{5.7}$$

$$\frac{V^{app}}{K_m^{app}} = \frac{Vb}{K_i^A K_m^B + K_m^A b} = \frac{\dfrac{V}{K_m^A} \cdot b}{\dfrac{K_i^A K_m^B}{K_m^A} + b} \tag{5.8}$$

In a typical experiment, various values of b would be used, and at each value of b the velocity would be determined at various values of a. Then V^{app} and K_m^{app} can be determined at each value of b exactly as in the one-substrate case, for example by a plot of a/v against a (cf. Section 2.5). Such a plot is called a *primary plot*, in order to distinguish it from the secondary plots that will be described shortly. *Figure 5.4* shows a typical set of primary plots for an

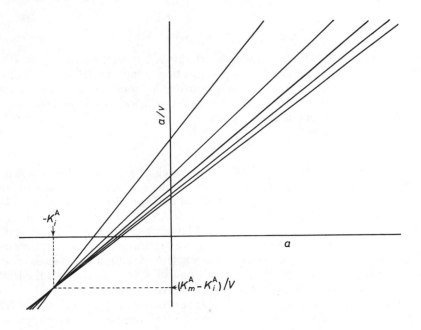

Figure 5.4 Primary plots of a/v against a at various values of b, for ternary-complex mechanisms, ignoring substrate inhibition: Plots of b/v against b at various values of a are similar

enzyme that obeys equation 5.5. It is characteristic of ternary complex mechanisms that the lines intersect at a point given by $a = -K_i^A$, $a/v = (K_m^A - K_i^A)/V$. It must occur to the left of the a/v axis, but can be either above or below the a axis, as K_m^A can be either greater or less than K_i^A.

Equations 5.6 and 5.8 are of the same form as the Michaelis–Menten equation, i.e. plots of V^{app} or V^{app}/K_m^{app} against a describe rectangular hyperbolas through the origin, and they can be analysed in exactly the same way. Thus equation 5.6 can be written as

84

$$\frac{b}{V^{\mathrm{app}}} = \frac{K_m^{\mathrm{B}}}{V} + \frac{1}{V} \cdot b$$

so that a *secondary plot* of b/V^{app} against b is a straight line of slope $1/V$ and intercept K_m^{B}/V on the b/V^{app} axis. Similarly, equation 5.8 gives

$$\frac{bK_m^{\mathrm{app}}}{V^{\mathrm{app}}} = \frac{K_i^{\mathrm{A}}K_m^{\mathrm{B}}}{V} + \frac{K_m^{\mathrm{A}}}{V} \cdot b$$

so that a secondary plot of $bK_m^{\mathrm{app}}/V^{\mathrm{app}}$ against b is a straight line of slope K_m^{A}/V and intercept $K_i^{\mathrm{A}}K_m^{\mathrm{B}}/V$ on the $bK_m^{\mathrm{app}}/V^{\mathrm{app}}$ axis. All four parameters. V, K_i^{A}, K_m^{A} and K_m^{B}, can readily be calculated from these plots, which are illustrated in *Figure 5.5*.

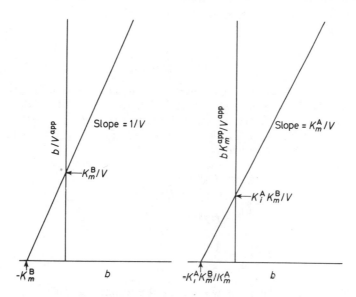

Figure 5.5 Secondary plots for ternary-complex mechanisms: The plot of b/V^{app} *against* b *is also applicable to substituted-enzyme mechanisms*

Equation 5.7 also describes a rectangular hyperbola, but the curve does not pass through the origin. Instead, $K_m^{\mathrm{app}} = K_i^{\mathrm{A}}$ when $b \longrightarrow 0$. It is thus a three-parameter hyperbola, and cannot be re-drawn as a straight line. As in other cases, K_m^{app} is a much less convenient parameter to examine than $K_m^{\mathrm{app}}/V^{\mathrm{app}}$.

One can equally well treat B as the variable substrate instead of A, making primary plots of b/v against b at the different values of a. The analysis is the same, and so there is no need to describe it again. The only important difference is that K_i^{B} does not occur in equation 5.5, and $K_i^{\mathrm{A}}K_m^{\mathrm{B}}/K_m^{\mathrm{A}}$ occurs wherever one might expect K_i^{B} from a simple interchange of A and B.

The primary and secondary plots are mainly useful in preliminary work and for illustrating results. For the definitive determination of kinetic

85

constants, it is preferable to use computational methods that permit all of the data to be analysed simultaneously. These methods are discussed in Chapter 10.

For the random-order ternary-complex mechanism, the initial rate in the absence of product is also given by equation 5.5 and so it is impossible to tell whether the order of binding is compulsory or random from initial velocity studies alone. Because the mechanism is symmetrical in A and B, $K_i^A K_m^B$ can be replaced with $K_i^B K_m^A$, and these two products are equal. If the equilibrium assumption is correct, i.e. if the breakdown of EAB to products is slow compared with the other first-order steps, then K_i^A and K_i^B are the dissociation constants of EA and EB, respectively, and K_m^A and K_m^B are the two dissociation constants of EAB, for loss of A and B, respectively. The graphical and computational analysis is the same as for the compulsory-order mechanism.

For the substituted-enzyme mechanism, the initial rate in the absence of product is given by

$$v = \frac{V ab}{K_m^B a + K_m^A b + ab} \tag{5.9}$$

The most striking feature of this equation is the absence of a constant from the denominator. (This was shown to be characteristic of all substituted-enzyme mechanisms in Section 3.7.) A distinctive pattern results if either substrate concentration is varied: for example, if a is varied, the apparent values of V and K_m are given by

$$V^{\mathrm{app}} = \frac{Vb}{K_m^B + b}$$

$$K_m^{\mathrm{app}} = \frac{K_m^A b}{K_m^B + b}$$

$$\frac{V^{\mathrm{app}}}{K_m^{\mathrm{app}}} = \frac{V}{K_m^A}$$

Only V^{app} varies in the same way as for the ternary-complex mechanisms. The important characteristic is that $V^{\mathrm{app}}/K_m^{\mathrm{app}}$ is constant, and equal to V/K_m^A. $V^{\mathrm{app}}/K_m^{\mathrm{app}}$ is also constant when b is varied, but is then equal to V/K_m^B. Primary plots of a/v against a or of b/v against b form a series of straight lines intersecting on the a/v or b/v axis, as shown in Figure 5.6. This pattern is readily distinguishable from the patterns of primary plots exhibited by the ternary-complex mechanisms (Figure 5.4) except in the rare case where K_i^A is much smaller than K_m^A.

The only secondary plot required for the substituted-enzyme mechanism is that of b/V^{app} against b, which has the same properties as the same plot for the ternary-complex mechanisms (Figure 5.5).

It is easy to understand why $V^{\mathrm{app}}/K_m^{\mathrm{app}}$ is independent of the concentration of the constant substrate when it is recalled (Section 2.3) that V/K_m is the pseudo-first-order rate constant for the reaction at very low substrate concentrations. If a approaches zero, the rate of production of E′ must be slow enough for B to be able to react with it as fast as it is formed, provided

86

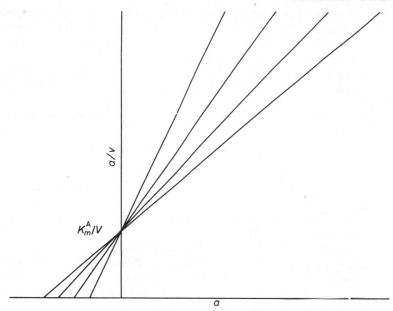

Figure 5.6 Primary plots of a/v *against* a *at various values of* b, *for substituted-enzyme mechanisms, ignoring substrate inhibition: Plots of* b/v *against* b *at various values of* a *are similar*

that *b* does not itself approach zero. As both product release steps are irreversible at initial time, the concentration of B can have no effect on the rate of the complete reaction under these conditions. In other words, V^{app}/K_m^{app} must be independent of *b*. By the same argument, V^{app}/K_m^{app} must be independent of *a* when *b* is varied. In contrast, in the compulsory-order ternary-complex mechanism, the two binding steps are not separated by an irreversible step: no matter how small *a* may be, the rate of the step EA + B ⇌ EAB remains dependent on *b*; and no matter how small *b* may be, the rate of this step remains dependent on the concentration of EA, which is itself dependent on *a*. Similar arguments apply to the random-order ternary-complex mechanism.

5.6 Substrate inhibition

The results in the previous section are strictly valid only at low substrate concentrations because, in all reasonable mechanisms, at least one of the four reactants can bind to the wrong enzyme species. In the substituted-enzyme mechanism, the substrate and product that lack the transferred group normally bind to the wrong form of the free enzyme; in the random-order ternary-complex mechanism, the same pair bind to the wrong binary complex; and in the compulsory-order ternary-complex mechanism, either the second substrate or the first product binds to the wrong binary complex. In this last case, substrate inhibition can occur in either the forward or the

87

reverse reaction, but not both, because only one of the two binary complexes is available. For convenience, we shall take B as the reactant that displays substrate inhibition for each mechanism, but the results can readily be transformed for other reactants if desired.

The non-productive complex EBQ in the compulsory-order ternary-complex mechanism was considered in Section 3.3. It can be allowed for in the rate equation by multiplying every term for EQ by $(1 + k_{+5}b/k_{-5})$, where k_{-5}/k_{+5} is the dissociation constant of EBQ. Hence equation 5.5 becomes

$$v = \frac{Vab}{K_i^A K_m^B + K_m^B a + K_m^A b + ab(1 + b/K_{si}^B)} \tag{5.10}$$

where K_{si}^B is *not* the same as k_{-5}/k_{+5} because the term in ab refers only partly to EQ. Depending on the relative amounts of (EAB–EPQ) and EQ at the steady state, K_{si}^B may approximate to k_{-5}/k_{+5}, or it may be much greater. Thus substrate inhibition may not be detectable with this mechanism at any attainable concentration of B.

Substrate inhibition according to equation 5.10 is effective only at high concentrations of A, and is thus uncompetitive. Primary plots of b/v against b are parabolic, with a common intersection point at $b = -K_i^A K_m^B/K_m^A$. Primary plots of a/v against a are linear, but have no common intersection point. These plots are illustrated in *Figure 5.7*.

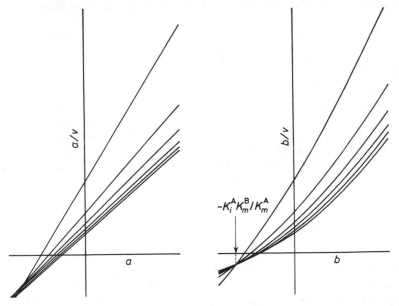

Figure 5.7 Effect of substrate inhibition by B (with $K_{si}^B = 10K_m^B$) on primary plots for ternary-complex mechanisms (cf. Figure 5.4)

In the random-order ternary-complex mechanism, the concentration of EQ is zero in the absence of added Q if the rapid-equilibrium assumption holds. As B cannot bind to a species that is absent, substrate inhibition does not occur with this mechanism unless Q is added. If the rapid-equilibrium assumption does not hold, there is no reason why substrate inhibition should not occur, but the form of it is difficult to predict with certainty because of the complexity of the rate equation. In this case, EBQ is *not* a dead-end complex (although some workers loosely refer to it as such) because it can be formed from either EB or EQ, and so it need not be in equilibrium with either.

In the substituted-enzyme mechanism, the non-productive complex EB results from the binding of B and E. It is a dead-end complex, and so $[EB]/[E] = b/K_{si}^B$, where K_{si}^B is the dissociation constant of EB. Equation 5.9 therefore becomes

$$v = \frac{Vab}{K_m^B a + K_m^A b(1 + b/K_{si}^B) + ab} \tag{5.11}$$

Inhibition according to this equation is most effective when a is small, and is thus competitive. Primary plots of b/v against b are again parabolic, but they intersect at $b = 0$, i.e. on the b/v axis. Primary plots of a/v against a are linear, with no common intersection point, but every pair of lines intersects at a positive value of a. These plots are illustrated in *Figure 5.8*.

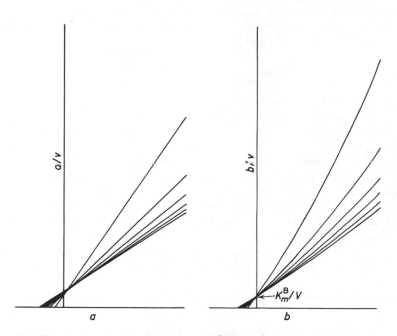

Figure 5.8 Effect of substrate inhibition by B (with $K_{si}^B = 10K_m^B$) on primary plots for substituted-enzyme mechanisms (cf. Figure 5.6)

Substrate inhibition might seem at first sight to be a tiresome complication in the analysis of kinetic data. Actually, it is very informative, because it accentuates the difference between ternary-complex and substituted-enzyme mechanisms, and is usually straightforward to interpret. As substrates normally bind better to the correct enzyme species than to the wrong species, substrate inhibition is rarely severe enough at low substrate concentrations to interfere with the analysis described in Section 5.5. Competitive substrate inhibition provides strong positive evidence for the substituted-enzyme mechanism. In contrast, the observation that primary plots intersect on the a/v or b/v axis is only negative evidence, as it can occur as a special case of a ternary-complex mechanism. Finally, when substrate inhibition occurs in the compulsory-order ternary-complex mechanism, it permits an identification of the second substrate, which would otherwise require product-inhibition studies.

5.7 Reverse reaction

If a reaction can be conveniently followed in the reverse direction, it is usually advantageous to do so in order to confirm and amplify the information gained about the mechanism from studying the forward reaction. For each of the group-transfer reactions considered, the reverse reaction is exactly analogous to the forward reaction. For example, omitting the terms in a and b from the complete rate equation for the compulsory-order ternary-complex mechanism (equation 5.2) yields

$$v = \frac{-V^r pq}{K_m^P K_i^Q + K_m^Q p + K_m^P q + pq} \tag{5.12}$$

which can be compared with equation 5.5 for the forward reaction. The minus sign merely indicates that the equation refers to the reverse reaction and can be omitted unless one wishes to consider the complete equation with all four reactants present.

Just as for single-substrate single-product reactions (Section 2.6), the kinetic constants for the forward and reverse directions of a multiple-substrate reaction are related to the equilibrium constant K by Haldane relationships. For any mechanism, the velocity at equilibrium must be zero, which in practice means that the numerator of the rate equation must be zero. So for equation 5.2,

$$\frac{V^f a_\infty b_\infty}{K_i^A K_m^B} - \frac{V^r p_\infty q_\infty}{K_m^P K_i^Q} = 0$$

i.e.

$$K = \frac{p_\infty q_\infty}{a_\infty b_\infty} = \frac{V^f K_m^P K_i^Q}{V^r K_i^A K_m^B} \tag{5.13}$$

where a_∞, b_∞, etc., represent the concentrations after infinite time, i.e. at equilibrium.

A second Haldane relationship can be found by rearranging the definitions of the kinetic constants shown in *Table 5.1*:

$$K = \left(\frac{V^f}{V^r}\right)^2 \frac{K_i^P K_m^Q}{K_m^A K_i^B} \tag{5.14}$$

Equation 5.13 does not depend on the definitions of the kinetic constants, and so is obviously true for any mechanism described by equation 5.2, i.e. for mechanisms that include isomerizations of EA, EAB or EQ. Perhaps surprisingly, equation 5.14 is also general and so applies to mechanisms that include isomerizations. This is not obvious and is not easy to prove, but it is true, and results from the fact that not only is the net velocity zero at equilibrium, but every individual step is also at equilibrium when the complete reaction is at equilibrium.

For any mechanism, one Haldane relationship can always be found to express the requirement of zero velocity at equilibrium, and additional relationships usually result from considering the definitions of the kinetic parameters in terms of rate constants. Thus the substituted-enzyme mechanism gives

$$K = \frac{K_i^P K_i^Q}{K_i^A K_i^B} = \frac{V^f K_i^P K_m^Q}{V^r K_i^A K_m^B} = \frac{V^f K_m^P K_i^Q}{V^r K_m^A K_i^B} = \left(\frac{V^f}{V^r}\right)^2 \frac{K_m^P K_m^Q}{K_m^A K_m^B}$$

and similar relationships exist for other mechanisms. Although these relationships can, in principle, be used to distinguish between mechanisms, it is difficult to do so in practice because the kinetic constants cannot usually be measured with sufficient accuracy to give an unambiguous answer. A more practical use of Haldane relationships is to check the self-consistency of results obtained in other ways. If all of the kinetic constants can be estimated, it is very easy to test the Haldane relationships, and it is worthwhile to do so routinely in order to round off a series of experiments.

Certain additional relationships can be found between the parameters of a rate equation, which can be valuable in favourable cases as they provide the only steady-state method for detecting isomerizations of intermediates. Such isomerizations do not affect the form of the rate equation (cf. Section 3.7), but they may remove constraints on the values that the parameters may have. For example, if the compulsory-order ternary-complex mechanism applies with no isomerizations, then the definitions shown in *Table 5.1* can be rearranged to give

$$\frac{V^f K_i^A}{V^r K_m^A} = 1 + \frac{k_{-1}}{k_{-2}}$$

and, as k_{-1} and k_{-2} must both be positive, it follows that

$$\frac{V^f K_i^A}{V^r K_m^A} \geqslant 1 \tag{5.15}$$

and, by a similar argument,

$$\frac{V^r K_i^Q}{V^f K_m^Q} \geqslant 1 \tag{5.16}$$

Unlike the Haldane relationships, these results are dependent on the definitions of the kinetic parameters, and are not necessarily true for mechanisms with isomerization steps: if EA isomerizes, equation 5.15 no longer applies, while if EQ isomerizes, equation 5.16 no longer applies. Thus the failure of either of these equations to hold experimentally provides evidence of isomerization of EA or EQ; but the converse does *not* hold, i.e. obedience to equations 5.15 and 5.16 does not rule out isomerization of EA or EQ. Moreover, this type of analysis does not indicate how many steps there may be in any isomerization, nor can isomerization of the ternary complex itself be detected. (For this information, measurements in the transient phase of the reaction are required, as is described in Chapter 9.) Nonetheless, this type of argument has been used to demonstrate the existence of isomerizations for potato phosphatase (Hsu, Cleland and Anderson, 1966) and a number of other enzymes.

5.8 Product inhibition

Of all the techniques that are available for elucidating reaction pathways, product-inhibition studies are among the most useful, being both informative and simple to understand and use. Provided that only one product is added to a reaction mixture, the term in the rate equation for the back reaction is zero (except for one-product reactions, which are rare, and are discussed in Section 5.10). The only effect of product addition, therefore, is to increase the denominator of the rate equation, i.e. to inhibit the reaction. The question of whether a product acts as a competitive, uncompetitive or mixed inhibitor cannot be answered in an absolute sense, because the answer depends upon which substrate is considered to be variable. However, once this has been decided, the question is very straightforward: the denominator of any rate equation can be divided into 'variable' and 'constant' terms, according to whether they contain the variable substrate concentration or not; the expression for V^{app} depends on the variable terms, while the expression for V^{app}/K_m^{app} depends on the constant terms, as shown in Section 5.5. So, if we recall the discussion of inhibitor types in Chapter 4, a product is a competitive inhibitor if its concentration appears only in constant terms, an uncompetitive inhibitor if it appears only in variable terms and a mixed inhibitor if it appears in both. If the product can combine with only one form of the enzyme, only linear terms in its concentration are possible, and so the inhibition is linear, but non-linear inhibition is also possible if the product can also bind to 'wrong' enzyme forms to give dead-end complexes.

Application of these principles to the equation for the compulsory-order ternary-complex mechanism (equation 5.2) shows that P is a mixed inhibitor whether A or B is the variable substrate, because the term in p is a 'constant' term and the term in abp is a 'variable' term for both substrates. On the other hand, q appears as a product with b, but not with a, and so Q is a competitive inhibitor with A as variable substrate, but a mixed inhibitor with B. The results for the back reaction are complementary: A is mixed for P, but competitive for Q, while B is mixed for both P and Q. All of these inhibitions are linear provided that the dead-end complex EBQ can be ignored. The occurrence of this complex causes terms in b^2q and ab^2 to appear in the rate equation (cf. Sections 3.3 and 3.7), with the result that inhibition by B in the back reaction becomes non-linear.

Certain types of product inhibition are eliminated when the constant substrate is present at a saturating concentration. For example, if A is saturating, terms in equation 5.2 that do not contain a become insignificant and q disappears from the rate equation, so that Q ceases to be an inhibitor when b is varied. On the other hand, p remains in both the 'constant' and 'variable' parts of the denominator, and so the type of inhibition is unchanged. Conversely, if B is saturating and a is varied, P becomes an uncompetitive inhibitor while Q remains a competitive inhibitor.

It is a simple matter to predict the product-inhibition characteristics of any other mechanism. As no new principles are required for the other two-substrate two-product mechanisms, the derivation of the results for these will be left as an exercise. The most reliable method is to examine the form of the rate equation, but the same results can usually be obtained by considering the mechanism without deriving the equation. Competitive inhibition can arise in either of two ways: if the inhibitor binds to the same species as the variable substrate in such a way that each excludes the other; or if it displaces the variable substrate when it binds (as in the Theorell–Chance mechanism, for example). Both possibilities mean that binding of the inhibitor prevents the substrate from binding, and have the same effect on the rate equation. Uncompetitive inhibition occurs if there is no reversible pathway between the binding of substrate and the binding of product. In two-product mechanisms, uncompetitive inhibition is largely confined to the case already considered, inhibition by the first product in the compulsory-order ternary-complex mechanism when the second substrate is saturating. However, in reactions with three or more products, uncompetitive inhibition is more common, and invariably occurs with at least one product in compulsory-order mechanisms. Consider, for example, the following mechanism:

$$
\begin{array}{ccccccc}
 & R & & & A & & \\
ER & \longleftarrow & E & \longleftarrow & & \longrightarrow & EA \\
\uparrow\; Q & & & & & & \uparrow\; B \\
\downarrow & & & & & & \downarrow \\
EQR & \longrightarrow & E' & \longleftarrow & & \longrightarrow & EAB \\
E'C & & & C & & P & E'P \\
\end{array}
$$

If only one product is present, steps in which other products are released are irreversible. Thus Q is an uncompetitive inhibitor for either A or B as variable substrate; R is uncompetitive for C; but P is not uncompetitive for any substrate. P becomes uncompetitive with A, however, if B is made saturating, because then the binding of B to EA becomes irreversible.

If a complete reaction is experimentally reversible, its mechanism cannot contain any inherently irreversible steps. However, steps can become irreversible in the two ways mentioned, either because a substrate is saturating or because a product is absent. This statement must be interpreted with caution, however. A saturating concentration is one that approaches infinity, not merely a large one, and the concentration required to make a step irreversible may be much higher than can be achieved experimentally. A similar caution applies to the absence of product; in some cases, particularly if the reaction is being studied in the thermodynamically unfavourable direction, the product may be such a potent inhibitor that its effects are noticeable in the very early part of the reaction. Fumarase provides a good example of this problem: Alberty *et al.* (1954) found that it was necessary to follow only the first few per cent of reaction in order to obtain even approximately constant initial rates. This was true at low substrate concentrations for both the forward and reverse reactions.

5.9 Isotope exchange

Study of the initial rates of multiple substrate reactions in both directions, and in the presence and absence of products, will usually eliminate many possible reaction pathways and give a good indication of the gross features of the mechanism, but it will usually not reveal the existence of any minor alternative pathways because these may contribute so little to the total rate that they are virtually undetectable. Further information is therefore required in order to provide a definitive picture. Even if a clear mechanism does emerge from initial-rate and product-inhibition experiments, it is valuable to be able to confirm its validity independently. The very important technique of isotope exchange, which was introduced to enzyme kinetics by Boyer (1959), can often satisfy these requirements.

In order to apply the results of isotope-exchange experiments, one must normally make two important assumptions. These are usually true and are often merely implied, but it is as well to state them clearly in order to prevent misunderstanding. The first assumption is that a reaction that involves radioactive substrates follows the same mechanism as the normal reaction, with the same rate constants. In other words, isotope effects are assumed to be negligible. This assumption is generally true, provided that tritium is not used as a radioactive atom. Even then, isotope effects are likely to be negligible if the tritium atom is not directly involved in the reaction or in binding the substrate to the enzyme. The second assumption is that the concentrations of all radioactive species are so low that they have no perceptible effect on the concentrations of unlabelled species. This assumption can usually be made to be true, and is very important, because it makes the analysis of results much simpler than it would otherwise be.

Isotope exchange can most readily be understood in relation to an example, such as the transfer of a radioactive atom (represented by an asterisk) from A* to P* in the compulsory-order ternary-complex mechanism:

94

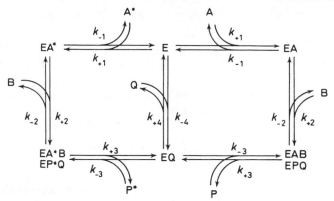

As this exchange requires the binding of A* to E, it can occur only if there is a significant concentration of E. Clearly, therefore, the exchange reaction will be inhibited by a high concentration of either A or Q, as they compete with A* for E. The effects of B and P are more subtle: on the one hand, the exchange reaction includes the binding of B to EA*, and so a finite concentration of B is required. On the other hand, if B and P are present at very high concentrations, the enzyme will exist largely as (EAB–EPQ) and so there will be no E to which A* can bind. One would therefore expect high concentrations of B and P to inhibit the exchange and it is not difficult to show that this expectation is correct. The rates of change of labelled intermediate concentrations can be written in the usual way, and set to zero according to the steady-state assumption:

$$\frac{d}{dt}[EA^*] = k_{+1}[E]a^* - (k_{-1} + k_{+2}b)[EA^*] + k_{-2}[EAB] = 0$$

$$\frac{d}{dt}[EA^*B] = k_{+2}b[EA^*] - (k_{-2} + k_{+3})[EA^*B] + k_{-3}p^*[EQ] = 0$$

Putting $p^* = 0$ and solving for $[EA^*B]$, we obtain

$$[EA^*B] = \frac{k_{+1}k_{+2}[E]a^*b}{k_{-1}(k_{-2} + k_{+3}) + k_{+2}k_{+3}b}$$

The initial rate of exchange, v^*, is given by $k_{+3}[EA^*B]$, or

$$v^* = \frac{k_{+1}k_{+2}k_{+3}[E]a^*b}{k_{-1}(k_{-2} + k_{+3}) + k_{+2}k_{+3}b} \tag{5.17}$$

In order to use this expression, $[E]$ must be known, but this presents no problem if the kinetic constants for the unlabelled reaction have been determined. It is simplest (and most usual) to study isotope exchange under conditions such that the unlabelled reactants are at equilibrium. In this case, $[E]$ is given by

$$[E] = \frac{e_0}{1 + \dfrac{k_{+1}a}{k_{-1}} + \dfrac{k_{+1}k_{+2}ab}{k_{-1}k_{-2}} + \dfrac{k_{-4}q}{k_{+4}}} \tag{5.18}$$

This expression can be substituted into equation 5.17 to give

$$v^* = \frac{k_{+1}k_{+2}k_{+3}e_0 a^* b}{\left(1 + \dfrac{k_{+1}a}{k_{-1}} + \dfrac{k_{+1}k_{+2}ab}{k_{-1}k_{-2}} + \dfrac{k_{-4}q}{k_{+4}}\right)[k_{-1}(k_{-2}+k_{+3})+k_{+2}k_{+3}b]} \tag{5.19}$$

p is not included in this equation because, if equilibrium is to be maintained, only three of the four reactant concentrations can be chosen at will. Any one of a, b and q can be replaced with p by means of the identity

$$K = \frac{k_{+1}k_{+2}k_{+3}k_{+4}}{k_{-1}k_{-2}k_{-3}k_{-4}} = \frac{pq}{ab}$$

If b and p are varied in a constant ratio (in order to maintain equilibrium) at fixed values of a and q, the effect on the exchange rate can be seen by realizing that the denominator of equation 5.19 is a quadratic in b, whereas the numerator is directly proportional to b. Hence the equation is of the same form as the equation for simple substrate inhibition (cf. Section 4.9). Therefore, as b and p are increased from zero to saturation, the exchange rate increases to a maximum and then decreases to zero.

The equations for any other exchange reaction can be derived in a similar manner. It is not necessary to maintain the unlabelled reactants at equilibrium, although it is much simpler to do so, because equilibrium equations are much simpler than steady-state equations. In the example we have considered, if equilibrium were not maintained, equation 5.18 would have to be replaced with the corresponding steady-state expression, i.e. equation 3.3, which is much less simple.

In the compulsory-order ternary-complex mechanism, exchange from B* to P* or Q* is not inhibited by A, because saturating concentrations of A do not remove EA, but in fact increase its concentration. Similar results apply to the reverse reaction, and exchange from Q* is inhibited by excess of P, but exchange from P* is not inhibited by excess of Q.

The random-order mechanism is distinguished from the compulsory-order mechanism by the fact that no exchange can be completely inhibited by the alternate substrate. For example, if B is present in excess, A* cannot bind to E but it can bind to EB instead, to give EA*B, which can break down to P* or Q*. As radioactive counting can be made very sensitive, it is possible to detect very minor alternative pathways by isotope exchange. The caution must be made, however, that isotope-exchange experiments require more highly purified enzyme than conventional kinetic experiments if valid results are to be obtained. The reason for this requirement is very simple. Suppose one is studying alcohol dehydrogenase, which catalyses the reaction

$$\text{ethanol} + \text{NAD}^+ \rightleftarrows \text{acetaldehyde} + \text{NADH}$$

A small amount of contaminating enzymes is of little importance if one is following the complete reaction, because it is unlikely that any of the contaminants is a catalyst for the complete reaction. However, exchange between NAD^+ and NADH is another matter: there are numerous enzymes that can catalyse this exchange, and one must therefore be certain that they are absent if one wants to obtain valid information about alcohol dehydrogenase.

Isotopic exchange permits a useful simplification of the substituted-enzyme mechanism, in that one can study one half of the reaction only:

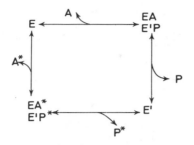

This mechanism is of the same form as the complete mechanism, with P* and A* replacing B and Q, respectively, but the kinetics are simpler because the rate constants are the same for the two halves of the reaction. This, of course, also provides an important qualitative distinction between the substituted-enzyme and ternary-complex mechanisms, as in ternary-complex mechanisms no exchange can occur unless the system is complete. This method of distinguishing between two types of mechanism was, in fact, used and discussed (Doudoroff, Barker and Hassid, 1947; Koshland, 1955) well before the introduction of isotope exchange as a kinetic technique.

The possibility of studying only parts of mechanisms in this way is particularly valuable with more complex substituted-enzyme mechanisms, with three or more substrates. In such cases, any simplification of the kinetics is obviously to be welcomed, and this approach has been used with some success, e.g. by Cedar and Schwartz (1969) in the study of asparagine synthetase.

5.10 Induced transport

Britton (1966, 1973) has introduced a very different application of isotope exchange, known as induced transport, that is very helpful in identifying enzyme isomerization steps in one-product reactions. In most respects, one-substrate, one-product reactions are much simpler than multiple-reactant mechanisms, but they possess the serious complication that one cannot study product inhibition without allowing for the back reaction. This makes the unequivocal identification of an enzyme isomerization very difficult. Now, the very nature of one-substrate one-product reactions, which are inevitably isomerizations, makes the occurrence of enzyme isomerization likely. The simplest way one can conceive of an enzyme catalysing, say, the interconversion of glucose 1-phosphate (GP) and glucose 6-phosphate (PG) is as a phosphoenzyme, E, that reacts to form a different phosphoenzyme, E', with a third step in which E' isomerizes back into E. The complete mechanism can be represented as shown in scheme 1. This is not, of course, the only possibility, and a second is suggested by the observation that phospho-glucomutase from rabbit muscle requires the presence of catalytic amounts of glucose 1,6-diphosphate (PGP). This can be accounted for within *Scheme 1* by assuming that the PGP is needed in order to convert a dephosphorylated

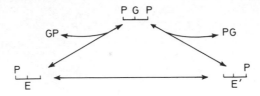

form of the enzyme into the active enzyme. But another possibility is shown in *Scheme 2*, where the enzyme operates by a ternary-complex mechanism in which PGP is both second substrate and first product.

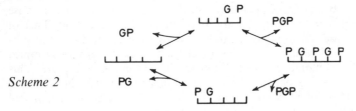

Scheme 2

Other mechanisms are also possible, but two will suffice for illustration. The important difference is that *Scheme 1* includes an enzyme isomerization whereas *Scheme 2* does not. The two mechanisms therefore ought to be readily distinguishable by product-inhibition experiments: if *Scheme 1* is correct, PG and GP should be mixed inhibitors; if *Scheme 2* is correct, they should be competitive with one another. Accordingly, Ray and Roscelli (1964) studied the reaction catalysed by phosphoglucomutase from rabbit muscle, and found the inhibition by both glucose phosphates to be purely competitive, with no detectable uncompetitive component. They concluded that either there was no enzyme isomerization or it was so fast that E and E′ could be regarded as a single species. Nevertheless, Britton and Clarke (1968) were able to show unequivocally that *Scheme 1* was correct, in an elegant series of experiments that are outlined below.

Britton and Clarke followed the exchange of radioactive label between GP and PG, but their technique differed from the usual one in the important respect that they allowed the labelled compounds to reach equilibrium before adding a large excess of unlabelled reactant, either GP or PG. They then followed the progress of the labelled reaction *away from* equilibrium. This may seem contrary to the laws of thermodynamics, but is in fact possible because of the large amount of free energy being dissipated in the unlabelled reaction. With ^{14}C-labelled reactants, G*P or PG*, they found that the labelled reaction proceeded in the opposite direction to the unlabelled reaction, an observation that is readily explained by *Scheme 1*: in order to be converted into PG*, G*P must first bind to E, but in the presence of a large amount of GP the concentration of E is negligible. However, even if k_{-3} and k_{+3} are large, a small amount of E' will always be available for PG* to bind to. Thus the reaction G*P → PG* is prevented by the lack of E, but the reverse reaction PG* → G*P is permitted by the presence of E'. Put another way, although there may initially be equilibrium between G*P and PG*, the enzymic reaction, represented by

$$E + G^*P \rightleftarrows E' + PG^*$$

is far from equilibrium because of the imbalance between E and E' produced by the high flux of unlabelled reactants. Hence a slow conversion of PG* into G*P should accompany the rapid conversion of GP into PG. Conversely, in the presence of a large excess of PG, the labelled reaction should proceed slowly from G*P to PG*.

In *Scheme 2*, ^{14}C exchange requires the participation of the unlabelled reactants, as follows:

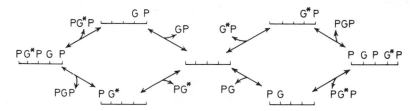

In the presence of a large excess of GP, the right-hand cycle can only proceed clockwise, while the left-hand cycle can only proceed anticlockwise. Thus transfer from G*P to PG* is possible, but the reverse is not. Hence, if this scheme is correct, transfer of label should proceed in the same direction as the unlabelled reaction, which is the contrary of what was observed.

Britton and Clarke were able to confirm the correctness of *Scheme 1* by studying ^{32}P-labelled reactants, GP* and P*G. They found that transfer was very slow and proceeded in the same direction as the unlabelled reaction. With *Scheme 1*, transfer of label in the phosphate group requires a six-step reaction, including ^{32}P-labelled enzyme as an intermediate:

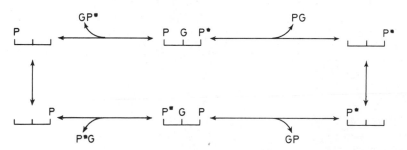

Now, in this scheme (reading clockwise), unlabelled PG must be released and unlabelled GP must be bound. Both of these reactions are facilitated by a large excess of GP, and the reverse reactions are prevented. Thus transfer

99

from GP* to P*G is possible, although not very favourable, because of a lack of free enzyme, whereas the reverse reaction is impossible. ^{32}P transfer should therefore proceed very slowly in the same direction as the unlabelled reaction, as observed.

In *Scheme 2*, transfer of ^{32}P requires a complex arrangement of three cycles, but leads to the same qualitative conclusion, i.e. transfer should proceed in the same direction as the unlabelled reaction. However, the kinetics are different for the two schemes and Britton and Clarke found that their observations were consistent only with *Scheme 1*.

In order to avoid unnecessary complication, induced transport has been discussed in a purely qualitative way, but the appropriate kinetic equations can be derived simply using an approach similar to that used for normal isotope exchange. Induced transport has not so far been widely applied, but the example of phosphoglucomutase should demonstrate the value of the technique in detecting features of mechanisms that product inhibition fails to detect.

6

Effects of pH and Temperature on Enzymes

6.1 pH and enzyme kinetics

Of the many problems that beset the earliest investigators of enzyme kinetics, none was more important than the lack of understanding of hydrogen-ion concentration, $[H^+]$. In aqueous chemistry, $[H^+]$ varies from about 1 M to about 10^{-14} M, which is an enormous range that is commonly reduced to more manageable proportions by the use of a logarithmic scale, $pH = -\log[H^+]$. All enzymes are profoundly influenced by pH, and no substantial progress could be made in the understanding of enzymes until Michaelis and his collaborators made pH control a routine characteristic of all serious enzyme studies. The stage had been set a few years earlier by Sørensen (1909), who had introduced the pH scale and described the use of buffers in a classic paper on the importance of the hydrogen-ion concentration in enzymic studies. Whatever doubts there may now be about the proper interpretation of pH effects in enzyme kinetics, the practical importance of pH continues undiminished: it is hopeless to attempt any kinetic studies without adequate control of pH.

It is perhaps surprising that it was left to an enzymologist to introduce such a generally useful term as pH, and it is worthwhile to reflect on the special properties of enzymes that created the need for it before it had been required in the much more highly developed science of chemical kinetics. With a few exceptions, such as pepsin and alkaline phosphatase, the enzymes that have been most studied are active only in aqueous solution at pH values in the range 5–9. Indeed, only pepsin has a physiologically significant activity outside this middle range of pH. Now, in the pH range 5–9, the hydrogen- and hydroxide-ion concentrations are in the range 10^{-5}–10^{-9} M, i.e. very low, and are very sensitive to impurities. Whole cell extracts, and crude enzyme preparations in general, are well buffered by enzyme and other polyelectrolyte impurities, but this natural buffering is lost when an enzyme is purified, and must be replaced with artificial buffers. Until this effect was realized, little progress in enzyme kinetics was possible. This situation can be contrasted with the situation in general chemistry: only a minority of reactions are studied in aqueous solution and, of these, the majority are studied either at very low or very high pH, at which the concentration of either hydrogen or hydroxide ion is high enough to be reasonably stable. Consequently, the early development of chemical kinetics was little hampered by the lack of understanding of pH.

The simplest type of pH effect on an enzyme, when only a single acidic or basic group is involved, is no different from the general case of hyperbolic inhibition and activation that was considered in Chapter 4. Conceptually, the protonation of a basic group on an enzyme is simply a special case of the binding of a modifier at a specific site and there is therefore no need to repeat the algebra for this simplest case. However, there are several differences between protons and other modifiers that make it worthwhile to examine protons separately. Firstly, virtually all enzymes are affected by protons, so that the proton is far more important than any other modifier. It is far smaller than any other chemical species and has no steric effect; this means that certain phenomena, such as pure non-competitive inhibition, are common with the proton as inhibitor but very rare otherwise. The proton concentration can be measured and controlled over a range that is enormously greater than that available for any other modifier and therefore one can expect to be able to observe any effects that might exist. Finally, protons normally bind to many sites on an enzyme, so that it is often insufficient to consider binding at one site only.

6.2 Ionization of a dibasic acid

Every enzyme contains a large number of acidic and basic groups. Of these groups, most are either fully deprotonated (aspartate and glutamate) or fully protonated (arginine and lysine) in the neutral pH range. However, there are always several groups with pK_a values in the range 5–9, notably the imidazole group of histidine and the sulphydryl group of cysteine, but also N-terminal amino groups and certain other groups when 'perturbed.' Hence there are inevitably several groups that change their state of ionization when the pH is varied, and one might therefore expect the treatment of enzyme ionization to be correspondingly complicated. Fortunately, however, the pH behaviour of many enzymes can be interpreted as a first approximation in terms of a simple model, due to Michaelis (1926), in which only two ionizable groups are considered. The enzyme may be represented as a dibasic acid, HEH, with two non-identical acidic groups:

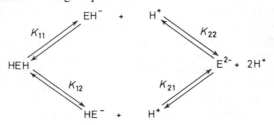

With the dissociation constants defined as shown in this scheme, the concentrations of all forms of the enzyme can be represented at equilibrium in terms of the hydrogen-ion concentration, $[H^+]$, or, more conveniently, h:

$$[EH^-] = [HEH]K_{11}/h$$
$$[HE^-] = [HEH]K_{12}/h$$
$$[E^{2-}] = [HEH]K_{11}K_{22}/h^2 = [HEH]K_{12}K_{21}/h^2 \tag{6.1}$$

Two points should be noted about these relationships: firstly, although K_{11} and K_{21} both define the dissociation of a proton from the same group, HE^- is more negative than HEH by one unit of charge and so one would expect it to be less acidic, i.e. $K_{11} > K_{21}$, not $K_{11} = K_{21}$; similarly, $K_{12} > K_{22}$. Secondly, the concentration of E^{2-} must (by the second law of thermodynamics) be the same whether it is derived from HEH via EH^- or via HE^-; the two expressions for $[E^{2-}]$ in equation 6.1 must therefore be equivalent, i.e. $K_{11}K_{22} = K_{12}K_{21}$.

If the total enzyme concentration is $e_0 = [HEH]+[EH^-]+[HE^-]+[E^-]$, then

$$[HEH] = \frac{e_0}{1 + \dfrac{K_{11}+K_{12}}{h} + \dfrac{K_{11}K_{22}}{h^2}} \tag{6.2}$$

$$[EH^-] = \frac{e_0 K_{11}/h}{1 + \dfrac{K_{11}+K_{12}}{h} + \dfrac{K_{11}K_{22}}{h^2}} \tag{6.3}$$

$$[HE^-] = \frac{e_0 K_{12}/h}{1 + \dfrac{K_{11}+K_{12}}{h} + \dfrac{K_{11}K_{22}}{h^2}} \tag{6.4}$$

$$[E^{2-}] = \frac{e_0 K_{11}K_{22}/h^2}{1 + \dfrac{K_{11}+K_{12}}{h} + \dfrac{K_{11}K_{22}}{h^2}} \tag{6.5}$$

These expressions show how the concentrations of the four species vary with h, and, by extension, with pH, and a typical set of curves is shown in *Figure 6.1*, with arbitrary values assumed for the dissociation constants. In a real experiment, one can never define the curves as precisely as this, because it is impossible to evaluate the four dissociation constants. The reason for this can be seen by considering the fact that $[EH^-]/[HE^-] = K_{11}/K_{12}$, i.e. a constant, independent of h. Thus no amount of variation of h will produce any change in $[EH^-]$ that is not accompanied by an exactly proportional change in $[HE^-]$. Consequently, it is impossible to determine how much of any given property is contributed by EH^- and how much by HE^- and for practical purposes we must therefore treat EH^- and HE^- as a single species, with concentration given by

$$[EH^-]+[HE^-] = e_0 / \left(\frac{h}{K_1} + 1 + \frac{K_2}{h} \right) \tag{6.6}$$

103

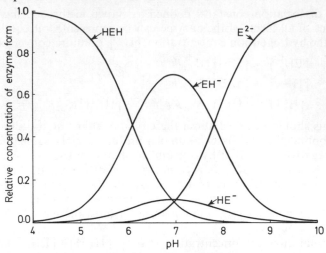

Figure 6.1 *Relative concentrations of enzyme forms as a function of pH, for an enzyme HEH with two ionizable groups: $pK_{11} = 6.1$; $pK_{12} = 6.9$; $pK_{21} = 7.0$; $pK_{22} = 7.8$*

where $K_1 = K_{11} + K_{12} = ([EH^-] + [HE^-])h/[HEH])$ and $K_2 = K_{11}K_{22}/(K_{11} + K_{12}) = [E^{2-}]h/([EH^-] + [HE^-])$. K_1 and K_2 are called *molecular dissociation constants*, to distinguish them from K_{11}, K_{12}, K_{21} and K_{22}, which are *group dissociation constants*. They have the practical advantage that they can be measured, whereas the conceptually preferable group dissociation constants cannot, because it is impossible to evaluate K_{12}/K_{11}.

The expressions for $[HEH]$ and $[E^{2-}]$ can also be written in terms of molecular dissociation constants:

$$[HEH] = e_0 \bigg/ \left(\frac{h^2}{K_1 K_2} + \frac{h}{K_2} + 1 \right)$$

$$[E^{2-}] = e_0 \bigg/ \left(1 + \frac{K_1}{h} + \frac{K_1 K_2}{h^2} \right)$$

We shall now examine equation 6.6 in more detail, because many enzymes display a 'bell-shaped' activity profile characteristic of this equation. There are several common misconceptions about this type of bell-shaped curve. Firstly, although it closely resembles a Gaussian curve from some values of pK_1 and pK_2 (i.e. $-\log K_1$ and $-\log K_2$), it is not, and has a noticeably flat maximum if $pK_2 - pK_1$ is greater than about 3. Secondly, the values of the pH when $[EH^-] + [HE^-]$ is half-maximal are not equal to pK_1 and pK_2, and are a poor approximation to them unless $pK_2 - pK_1$ is large (which it often is not). However, the mean of these two pH values *is* equal to $\frac{1}{2}(pK_1 + pK_2)$, and is also the pH at which the maximum occurs. The relationship between the half-width of the curve and $pK_2 - pK_1$ is shown in *Table 6.1*, and some representative bell-shaped curves are shown in *Figure 6.2*. Thirdly,

104

Table 6.1 RELATIONSHIP BETWEEN THE HALF-WIDTH* AND THE pK DIFFERENCE FOR BELL-SHAPED pH PROFILES

Half-width	$pK_2 - pK_1$	Half-width	$pK_2 - pK_1$	Half-width	$pK_2 - pK_1$
1.14†	$-\infty$	2.1	1.73	3.1	3.00
1.2	-1.27	2.2	1.88	3.2	3.11
1.3	-0.32	2.3	2.02	3.3	3.22
1.4	0.17	2.4	2.15	3.4	3.33
1.5	0.51	2.5	2.28	3.5	3.44
1.6	0.78	2.6	2.41	3.6	3.54
1.7	1.02	2.7	2.53	3.7	3.65
1.8	1.22	2.8	2.65	3.8	3.76
1.9	1.39	2.9	2.77	3.9	3.86
2.0	1.57	3.0	2.88	4.0	3.96

* The half-width is defined as the difference between the pH values at which the ordinate has half of its maximum value.
† The half-width cannot be less than 1.14 unless the pH profile derives from a more complex mechanism than that considered in the text, e.g. one that involves more than two ionizable groups.

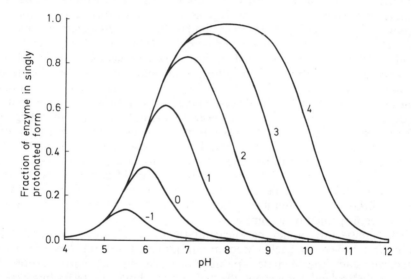

Figure 6.2 Bell-shaped curves calculated from equation 6.6 with $pK_1 = 6.0$ and $pK_2 = 5.0$–10.0: Each curve is labelled with the value of $pK_2 - pK_1$, as this quantity determines its shape

even supposing that pK_1 and pK_2 are correctly estimated, the values of the group dissociation constants remain unknown, unless plausibility arguments are invoked, with unprovable assumptions. Finally, although the condition $pK_2 < pK_1$ is not impossible, it is not common because it requires co-operative protonation (Dixon, 1973; *see also* Chapter 7), i.e. the group dissociation constants disobey the relationships $K_{11} > K_{21}$ and $K_{12} > K_{22}$. In fact, if these relationships hold, K_1 must be at least $4K_2$, i.e. $pK_2 - pK_1 \geqslant 0.6$.

105

6.3 Effect of pH on enzyme kinetic constants

By a simple extension of the theory for the ionization of a dibasic acid, one can account for the bell-shaped pH activity curves that are often observed for the enzyme kinetic constants V and V/K_m. (The treatment of K_m is more complex, as will be seen.) The basic mechanism is as follows:

$$
\begin{array}{ccc}
H_2E & & H_2ES \\
\updownarrow K_1^E & & \updownarrow K_1^{ES} \\
S + HE^- & \underset{k_{-1}}{\overset{k_{+1}}{\rightleftarrows}} HES^- & \overset{k_{+2}}{\longrightarrow} HE^- + P \\
\updownarrow K_2^E & & \updownarrow K_2^{ES} \\
E^{2-} & & ES^{2-}
\end{array}
$$

The free enzyme is again treated as a dibasic acid, H_2E, with two molecular dissociation constants, K_1^E and K_2^E, as in the previous section, and the enzyme–substrate complex H_2ES is similar, but with dissociation constants K_1^{ES} and K_2^{ES}. Only the singly ionized complex, HES^-, is able to react to give products. Before proceeding further, it must be emphasized that this scheme includes several implied assumptions that may be over-simplifications. Firstly, the omission of substrate-binding steps for H_2E and E^{2-} implies that the protonation steps are dead-end reactions, so that they can be treated as equilibria (*see* Section 3.6). However, this is nothing more than begging the question, because in most cases it is most unreasonable to postulate that S cannot bind directly to H_2E and E^{2-}. If these steps are included, the protonation steps cease to be dead-end reactions, and can then be treated as equilibria only if they are assumed to be very rapid compared with other steps. This may seem to be a reasonable assumption, in view of the simple nature of the reaction, but it may not always be true, particularly if protonation is accompanied by a compulsory conformation change.

The scheme also implies that the catalytic reaction involves only two steps, as in the simplest Michaelis–Menten mechanism. If several steps are postulated, with each intermediate capable of protonation and deprotonation, the form of the final equation is not affected, but any hope of interpreting experimental results in a straightforward way is lost. (Compare the effect of introducing an extra step into the simple Michaelis–Menten mechanism, Section 2.6.)

Finally, the assumption that only HES^- can break down to give products may not always be true, but it is likely to be reasonable for many enzymes because most enzyme activities do approach zero at high and low pH values. Moreover, it is widely believed that many enzymes owe their remarkable activity to the joint reactivity of an acidic and a basic group.

Recognizing that the present scheme may be an optimistic representation of the actual situation, let us consider the rate equation that it predicts. If there were no pH behaviour, i.e. if HE^- and HES^- were the only enzyme species, then the scheme would reduce to the ordinary Michaelis–Menten mechanism, with a rate given by

$$v = \frac{k_{+2}e_0 s}{\left(\dfrac{k_{-1}+k_{+2}}{k_{+1}}\right) + s} = \frac{\tilde{V}s}{\tilde{K}_m + s}$$

where $\tilde{V} = k_{+2}e_0$ and $\tilde{K}_m = (k_{-1}+k_{+2})/k_{+1}$ are the *pH-corrected* constants, convenient fictions analogous to the 'expected' values discussed in the context of non-productive binding (Section 4.8). In reality, however, the free enzyme does not exist solely as HE^-, nor the enzyme–substrate complex solely as HES^-. The full rate equation is

$$v = \frac{k_{+2}e_0 s}{\left(\dfrac{h}{K_1^E} + 1 + \dfrac{K_2^E}{h}\right)\dfrac{k_{-1}+k_{+2}}{k_{+1}} + \left(\dfrac{h}{K_1^{ES}} + 1 + \dfrac{K_2^{ES}}{h}\right)s}$$

which we can write as

$$v = Vs/(K_m + s)$$

where

$$V = \tilde{V} \left/ \left(\dfrac{h}{K_1^{ES}} + 1 + \dfrac{K_2^{ES}}{h}\right)\right. \tag{6.7}$$

$$K_m = \left(\dfrac{h}{K_1^E} + 1 + \dfrac{K_2^E}{h}\right)\tilde{K}_m \left/ \left(\dfrac{h}{K_1^{ES}} + 1 + \dfrac{K_2^{ES}}{h}\right)\right. \tag{6.8}$$

$$V/K_m = (\tilde{V}/\tilde{K}_m) \left/ \left(\dfrac{h}{K_1^E} + 1 + \dfrac{K_2^E}{h}\right)\right. \tag{6.9}$$

V reflects the ionization of the enzyme–substrate complex; V/K_m reflects the ionization of the free enzyme; while K_m, as usual, is complicated, being affected by both. For V or V/K_m, the dependence on pH follows a symmetrical bell-shaped curve, of exactly the form discussed in the previous section. For K_m, it is not generally profitable to attempt to analyse the pH behaviour unless it is very simple, as in the special case considered in the next section.

6.4 pH independence of K_m

It is common for K_m to be independent of pH in a pH range where V varies greatly. When this happens, it may be taken as evidence that K_m is a true equilibrium constant (Haldane, 1930), but the arguments that lead to this conclusion require careful examination as it can scarcely be claimed that it is obvious. In fact, examination of equation 6.8 suggests that the condition $K_1^E = K_1^{ES}$, $K_2^E = K_2^{ES}$ is sufficient for K_m to be independent of pH, regardless of the values of k_{-1} and k_{+2}. However, this is incorrect, because equation 6.8 was derived from an unrealistic model, as discussed in the previous section. As a remedy, steps for the binding of substrate to H_2E and E^{2-} must

107

be included. It is also worthwhile to separate binding steps from catalytic steps, i.e. to represent HES^- and HEP^- as distinct species. Then we can make the reasonable assumption that correct protonation is necessary for the second, catalytic, step but is of no importance in binding steps, i.e. protons have no effect on substrate or product binding, and *vice versa*. Then the model becomes

$$
\begin{array}{ccccccc}
S + H_2E & \underset{k_{-1}}{\overset{k_{+1}}{\rightleftharpoons}} & H_2ES & & H_2EP & \underset{k_{-3}}{\overset{k_{+3}}{\rightleftharpoons}} & H_2E + P \\
\updownarrow K_1 & & \updownarrow K_1 & & \updownarrow K_1 & & \updownarrow K_1 \\
S + HE^- & \underset{k_{-1}}{\overset{k_{+1}}{\rightleftharpoons}} & HES^- & \underset{k_{-2}}{\overset{k_{+2}}{\rightleftharpoons}} HEP^- & \underset{k_{-3}}{\overset{k_{+3}}{\rightleftharpoons}} & HE^- + P \\
\updownarrow K_2 & & \updownarrow K_2 & & \updownarrow K_2 & & \updownarrow K_2 \\
S + E^{2-} & \underset{k_{-1}}{\overset{k_{+1}}{\rightleftharpoons}} & ES^{2-} & & EP^{2-} & \underset{k_{-3}}{\overset{k_{+3}}{\rightleftharpoons}} & E^{2-} + P
\end{array}
$$

It is not easy to derive a steady-state rate equation for such a complex mechanism, even with the aid of the King–Altman method (Chapter 3), as there are no fewer than 384 patterns to be considered. Fortunately, the problem is greatly simplified by assuming that protonation and deprotonation steps are very fast compared with other steps and by making full use of the simplification described in Section 3.6. The rate equation turns out to be the Michaelis–Menten equation, with V and K_m defined by

$$
V = \frac{k_{+2}k_{+3}e_0 f(h)}{(k_{+2}+k_{-2})f(h)+k_{+3}} \tag{6.10}
$$

$$
K_m = \frac{1}{k_{+1}}\left[k_{-1} + \frac{k_{+2}(k_{+3}-k_{-1})f(h)}{(k_{+2}+k_{-2})f(h)+k_{+3}}\right] \tag{6.11}
$$

where $f(h)$ is the Michaelis function:

$$
f(h) = 1 \left(\frac{h}{K_1} + 1 + \frac{K_2}{h}\right) \tag{6.12}
$$

We must now examine equations 6.10 and 6.11 in order to determine the circumstances in which V can be pH dependent with K_m simultaneously independent of pH. There are in fact two such circumstances: either $k_{+3} = k_{-1}$, or k_{+2} is very small. In either case, K_m simplifies to k_{-1}/k_{+1}, i.e. the equilibrium dissociation constant of HES^-. Thus the original assertion is proved, provided that we accept that the model is reasonable. Actually, it is difficult to avoid reaching a similar conclusion even if the model is made more general: if more than three steps are included in the reaction, the algebra becomes more complicated but the conclusion is unaltered; if all of the acid dissociation constants for H_2E, H_2ES and H_2EP are assumed to be different, it becomes very difficult to make K_m independent of pH without introducing much less reasonable assumptions at the same time.

It is of interest that it is possible for K_m to be equal to k_{-1}/k_{+1} without the need for k_{+2} to be small, if $k_{-1} = k_{+3}$. At first sight, this appears to be such an unlikely coincidence that it can be dismissed from serious consideration. However, when it is recalled that in many metabolic reactions the substrate and product are structurally very similar to one another, and may be expected to bind to the enzyme in essentially the same way, it will be realized that k_{-1} and k_{+3} may easily be equal, at least approximately, as they are the rate constants for closely analogous reactions.

Finally, a careful comparison ought to be made between equations 6.7 and 6.10, as equation 6.7 is more convenient for application to experimental results whereas equation 6.10 is more likely to represent the true situation. Substituting equation 6.12 for $f(h)$ in equation 6.10 and rearranging, we obtain

$$V = \frac{\alpha k_{+2} e_0}{\dfrac{\alpha h}{K_1} + 1 + \dfrac{\alpha K_2}{h}}$$

where $\alpha = k_{+3}/(k_{+2} + k_{-2} + k_{+3})$. Comparing this equation with equation 6.7, we see that all of the experimentally measurable quantities $\tilde{V}$, K_1^{ES} and K_2^{ES} are perturbed from their theoretical values of $k_{+2}e_0$, K_1 and K_2, respectively, by the unknown factor α, i.e.

$$\tilde{V}/k_{+2}e_0 = K_1/K_1^{ES} = K_2^{ES}/K_2 = \alpha$$

From its definition, α cannot exceed unity, but nothing else can be assumed about its value. Further, as measured dissociation constants are always molecular rather than group dissociation constants, it is questionable whether any valid information can be obtained about the protonation equilibria of enzymes from kinetic measurements in the steady state. This scepticism appears to be fully justified, and it is wise to be very cautious when interpreting pH-dependence experiments. Nonetheless, almost all studies of the pH dependence of enzyme kinetics have been interpreted according to the model in Section 6.3, and measured dissociation constants have usually been regarded as group dissociation constants, but it is hard to find any clear evidence that seriously erroneous conclusions have been reached in any single example.

6.5 Ionization of groups remote from the active site

The pH behaviour of enzymes is almost always discussed in terms of a very small number of ionizing groups, often only one or two, yet all enzymes contain many more than two ionizing groups and one may wonder whether a simple treatment can have any real validity. Part of the answer to this difficulty has been given already, in Section 6.2: many groups ionize to a significant extent only outside the experimental pH range, and these groups do not complicate the picture. The remainder can be divided into three classes: those that are directly involved in the catalytic or binding activity of the enzyme; those that are indirectly involved, e.g. because they are close to the catalytic groups or because they are required in a particular ionic form in order to maintain the conformation of the enzyme molecule; and those that are remote from the active site and have no perceptible effect on the catalytic activity. Any reasonable model must allow for the ionization of groups in the first class, and usually for groups in the second class also. Provided that there are only a few of these groups, it may be possible to devise a model that is simple enough to be manageable. Groups in the third class can legitimately be ignored. This may seem intuitively to be obvious, but it is dangerous to regard any unproved assertion as obvious and it is worthwhile to demonstrate the truth of this particular one, at least for a very simple case.

Let us again consider the case of a dibasic acid, as in Section 6.2:

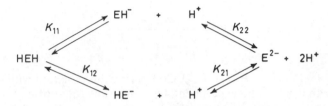

Let us also assume that the two groups ionize independently of one another, i.e. $K_{21} = K_{11}$ and $K_{22} = K_{12}$, and that only one of the two groups has any relevance to the activity of the enzyme, so that (for example) EH^- and E^{2-} are equally active and HEH and HE^- have no activity. In this case, the activity is proportional to $([EH^-]+[E^{2-}])$. Substituting $K_{21} = K_{11}$ and $K_{22} = K_{12}$ into equations 6.3 and 6.5, we obtain

$$[EH^-]+[E^{2-}] = e_0\left(\frac{K_{11}}{h} + \frac{K_{11}K_{12}}{h^2}\right)\bigg/\left(1 + \frac{K_{11}+K_{12}}{h} + \frac{K_{11}K_{12}}{h^2}\right)$$

$$= e_0\bigg/\left(\frac{h}{K_{11}} + 1\right)$$

Thus K_{12} cancels from the equation, and the pH dependence is seen to be exactly what it would be if the second group had not been considered. Similar simplifications occur in all models that allow for the ionization of groups that are irrelevant to the activity.

6.6 Change of rate-determining step with pH

The Michaelis treatment in terms of a dibasic acid is the most commonly invoked explanation of bell-shaped pH activity curves, but it is not only, or even the simplest, explanation of such effects. An alternative approach is to assume that there is only one ionizable group to be considered, and that it must be protonated in one step of the reaction but deprotonated in another, as in the following example:

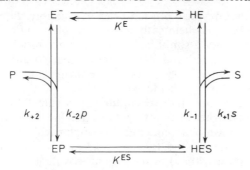

It is clear from inspecting the mechanism that, at very high pH values, when the concentration of HE is vanishingly small, substrate binding must be very slow and hence rate determining, but at very low pH values, the concentration of EP becomes vanishingly small and so product release must be rate determining. The velocity of the reaction must therefore display a bell-shaped dependence on pH, approaching zero at both extremes. A similar conclusion can be reached more rigorously by considering the form of the rate equation. If the protonation steps can be treated as equilibria, the rate obeys the Michaelis–Menten equation, with pH-dependent parameters as follows:

$$V = k_{+2}K^{ES}e_0/(K^{ES}+h)$$

$$K_m = \frac{(k_{-1}h+k_{+2}K^{ES})(K^E+h)}{k_{+1}h(K^{ES}+h)}$$

$$V/K_m = k_{+1}k_{+2}K^{ES}e_0h/(k_{-1}h+k_{+2}K^{ES})(K^E+h)$$

It can be seen that V displays a simple pH transition, with a single ionization constant equal to K^{ES}, the acid dissociation constant of HES. V/K_m displays a bell-shaped pH profile of the same shape as in the Michaelis treatment. K_m, as usual, is complex and it is not very useful to discuss it separately.

If the protonation steps are not treated as equilibria, it is still possible to derive a manageable rate equation for this mechanism, which is simply a compulsory-order ternary-complex mechanism with H^+ as both first substrate and first product. Thus the equation for that mechanism (equation 5.2, with appropriate changes in symbols) applies. This more rigorous treatment shows that V displays a bell-shaped profile as well as V/K_m.

6.7 Temperature dependence of enzyme-catalysed reactions

In principle, the theoretical treatment discussed in Sections 1.6 and 1.7 for the temperature dependence of simple chemical reactions applies equally well to enzyme-catalysed reactions, but in practice several complications arise that must be properly understood if any useful information is to be obtained from temperature-dependence studies. Firstly, almost all enzymes become denatured if they are heated much above physiological temperatures and the

111

conformation of the enzyme is altered, often irreversibly, with loss of catalytic activity. Denaturation is chemically a very complex and only partly understood process, and only a simplified account is given here. Only reversible denaturation is considered and it is assumed that equilibrium exists at all times between the active and denatured enzyme and that only a single denatured species need be considered.

Denaturation does not involve rupture of covalent bonds, but only of hydrogen bonds and other weak interactions that are involved in maintaining the active conformation of an enzyme. Although each individual bond is far weaker than a covalent bond (about 20 kJ mol^{-1} for a hydrogen bond, compared with about 400 kJ mol^{-1} for a covalent bond), denaturation generally involves the rupture of a large number of them. The standard enthalpy of reaction, $\Delta H°$, for denaturation is therefore often very high, typically 200–500 kJ mol^{-1}. However, the rupture of a large number of hydrogen bonds greatly increases the number of conformational states available to an enzyme molecule and so denaturation is also characterized by a very large standard entropy of reaction, $\Delta S°$.

The effect of denaturation on observed enzymic rate constants can be seen by considering a simple example:

$$\begin{array}{c} E' \\ K \Big\updownarrow \\ E + S \xrightarrow{k} E + P \end{array}$$

This scheme represents an active enzyme, E, in equilibrium with an inactive form, E′, and the catalytic reaction is treated as a simple second-order reaction, as is usually observed at very low substrate concentrations. The equilibrium constant, K, for denaturation varies with temperature according to the van't Hoff equation (Section 1.6):

$$-RT \ln K = \Delta G° = \Delta H° - T\Delta S°$$

where R is the gas constant, T is the absolute temperature and $\Delta G°$, $\Delta H°$ and $\Delta S°$ are the standard Gibbs free energy, enthalpy and entropy of reaction, respectively. This relationship can be re-written as an expression for K:

$$K = \exp(\Delta S°/R - \Delta H°/RT)$$

The rate constant k is governed by the integrated Arrhenius equation:

$$k = A \exp(-E_a/RT)$$

where A is a constant and E_a is the energy of activation. The rate of reaction is given by $v = k[E][S]$, but the active enzyme concentration, $[E]$, needs to be expressed in terms of the total enzyme concentration, e_0, and so

$$v = ke_0s/(1+K)$$

The observed rate constant, k^{obs}, may be defined as $k/(1+K)$, and varies with

112

temperature according to the equation

$$k^{obs} = \frac{A \exp(-E_a/RT)}{1 + \exp(\Delta S^\circ/R - \Delta H^\circ/RT)}$$

At low temperatures, when $\Delta S^\circ/R$ is small compared with $\Delta H^\circ/RT$, the exponential term in the denominator is insignificant, and so k^{obs} varies with temperature in the ordinary way according to the Arrhenius equation. However, at temperatures above $\Delta H^\circ/\Delta S^\circ$, the denominator increases very steeply with temperature and the velocity decreases rapidly to zero.

Although this model is over-simplified, it does show why the Arrhenius equation appears to fail for enzyme-catalysed reactions at high temperatures. In the older literature, it was common for *optimum temperatures* for enzymes to be reported, but the temperature at which k^{obs} is a maximum is of no particular significance, as the temperature dependence of enzyme-catalyzed reactions is often found in practice to vary with the experimental procedure. In particular, the longer reaction mixtures are incubated before analysis, the lower the 'optimum temperature' is found to be. The explanation of this effect is that denaturation often occurs fairly slowly, so that the reaction cannot properly be treated as an equilibrium. The extent of denaturation therefore increases with the time of incubation, which is not usually a problem with modern experimental techniques, because in continuously assayed reaction mixtures time-dependent processes are usually very obvious.

Because of denaturation, straightforward results can usually be obtained from temperature-dependence studies of enzymes only within a fairly small range of temperature, say between 0 and 50°C, but even within this range there are important hazards to be avoided when one is interpreting results. Very little significance can be attached to studies of the temperature dependence of V, V/K_m or K_m unless the mechanistic meanings of these parameters are known. Thus, if K_m is a function of several rate constants, its temperature dependence is likely to be complex combination of different effects, and of little theoretical significance or interest; but if K_m is known with reasonable certainty to be a true dissociation constant, its temperature dependence can provide useful thermodynamic information about the enzyme. Similarly, the temperature dependence of V is of interest only if there is some knowledge of the step in the mechanism to which it refers.

Another difficulty that must be borne in mind is that pH and temperature effects are not, in general, independent, because most ionization constants are temperature dependent. It is therefore likely to be difficult to interpret the results of temperature studies unless pH-corrected parameters (*see* Section 6.3) are estimated. In this connection, it is as well to remember that K_w, the ionization constant of water, is highly temperature dependent, and pK_w (i.e. $-\log K_w$) is 14.0 only at 24°C. At 37°C, where enzymic reactions are often studied, $pK_w = 13.62$, and so neutrality at 37°C occurs at pH 6.8 and not at pH 7.0. Much larger deviations occur at higher temperatures. This consideration is of the greatest importance in studies of reactions that involve OH^- ions directly, particularly if enzyme-catalysed rates are compared with base-catalysed rates.

6.8 Use of temperature for studying enzyme specificity

The tone of the previous section was intended to discourage slapdash studies of the temperature dependence of enzyme-catalysed reactions, but it would be wrong to suggest that no useful information can be obtained from temperature-dependence studies; on the contrary, if proper care is taken, very valuable information about enzyme reaction mechanisms can be obtained. Enzyme specificity is a particularly fruitful area of study. Provided that one can be sure that one is comparing like with like—for example k_{cat} (i.e. V/e_0) for one substrate must refer to the same step in the mechanism as k_{cat} for another substrate if the two are to be compared—then comparison of the activation parameters $\Delta H^{\ddagger}$ and $\Delta S^{\ddagger}$ is often much more informative than comparison of the simple rate constants. A classic study of this type was carried out by Bender, Kézdy and Gunter (1964) for the α-chymotrypsin-catalysed hydrolysis of numerous substrates. This enzyme is generally acknowledged to operate by an *acyl-enzyme* mechanism, i.e. a substituted-enzyme mechanism in which the substrate RCO–X first acylates the enzyme, HE:

$$\text{RCO–X} + \text{HE} \underset{k_{-1}}{\overset{k_{+1}}{\rightleftharpoons}} (\text{RCO–X·HE}) \xrightarrow{k_{+2}} \text{RCO–E} + \text{HX}$$

with the release of the first product HX, and the acyl-enzyme RCO–E then reacts with water to regenerate the free enzyme and release the second product, RCO–OH:

$$\text{RCO–E} + \text{H}_2\text{O} \xrightarrow{k_{+3}} \text{RCO–OH} + \text{HE}$$

As $k_{+2} \gg k_{+3}$ for some substrates and $k_{+2} \ll k_{+3}$ for others, direct comparison of k_{cat} values is likely to be meaningless, but by comparing the results for different substrates Bender, Kézdy and Gunter were able to determine the values of k_{+2} and k_{+3} separately. For example, a series of ester derivatives of acetyl-L-tyrosine share a common value of k_{cat} that is different from the value of k_{cat} found for a similar series of derivatives of acetyl-L-tryptophan. This suggests that $k_{+2} \gg k_{+3}$ and $k_{cat} = k_{+3}$ for these substrates. By this type of method, they were able to measure k_{+3} at different temperatures for a series of acylchymotrypsins. They found a wide variation between derivatives of specific substrates, such as acetyl-L-tyrosylchymotrypsin, and derivatives of non-specific substrates, such as acetylchymotrypsin. However, part of this variation may be due to differences in the inherent (non-enzymic) reactivity of the acyl groups, and to correct for this Bender, Kézdy and Gunter divided each value of k_{+3} by the rate constant for saponification (i.e. hydrolysis catalysed by hydroxide ion) of the corresponding ethyl esters. They then calculated $\Delta H^{\ddagger}$ and $\Delta S^{\ddagger}$ for the corrected values of k_{+3}, with the results shown in *Table 6.2*.

The most striking characteristic of these results is the absence of any significant correlation between k_{+3} and $\Delta H^{\ddagger}$. Indeed, $\Delta H^{\ddagger}$ varies very little, and it appears that energetic considerations are not of great importance in

Table 6.2 ACTIVATION PARAMETERS FOR THE HYDROLYSIS OF SOME ACYLCHYMOTRYPSINS (after Bender, Kézdy and Gunter, 1964)

Acyl group	Relative k_{+3} (corrected)	$\Delta H^{\ddagger}$ kJ mol^{-1}	$\Delta S^{\ddagger}$ J mol^{-1} K^{-1}
Acetyl-L-tyrosyl	3 540	43.3	− 56.3
Acetyl-L-tryptophanyl	942	50.4	− 83.2
trans-Cinnamoyl	15	47.0	− 124.3
Acetyl	1	40.7	− 150.8

this step of the reaction. Instead, the variation in k_{+3} is almost wholly the result of a wide variation in $\Delta S^{\ddagger}$. This implies that the rate of the reaction is determined mainly by the probability that the acyl group will adopt an appropriate orientation in the active site. For a bulky and specific group, such as the acyl-L-tyrosyl group, only a few orientations are possible, so that the correct one has a fairly high probability, but for a small and unspecific group, such as the acetyl group, very many orientations are possible, and the correct one has a much lower probability. This interpretation is supported by the fact that comparisons between acylchymotrypsins derived from L- and D-amino acids (Ingles and Knowles, 1967) led to very similar conclusions about the specificity of chymotrypsin, as discussed in Section 4.8.

7

Control of Enzyme Activity

7.1 Necessity for metabolic control

It is obvious that all living organisms require a high degree of control over metabolic processes so as to permit ordered change without precipitating catastrophic progress towards thermodynamic equilibrium. It is less obvious that enzymes that behave in the way described in other chapters are unlikely to be able to provide the necessary degree of control. Hence it is appropriate to begin by examining an important step in metabolism, the interconversion of fructose 6-phosphate (F6P) and fructose 1,6-diphosphate (FDP), with a view to defining the qualities that are needed in controlled enzymes. The conversion of F6P to FDP requires ATP:

$$F6P + ATP \rightarrow FDP + ADP$$

It is catalysed by phosphofructokinase and is the first step in glycolysis that is unique to glycolysis, i.e. the first step that does not form part of other metabolic processes as well. It is thus an appropriate step for the control of the whole process, and there is little doubt that it is indeed the major control point. Under metabolic conditions, the reaction is essentially irreversible and, in gluconeogenesis, it is by-passed by a hydrolytic reaction, catalysed by fructose diphosphatase:

$$FDP + H_2O \rightarrow F6P + phosphate$$

This reaction is also essentially irreversible. The parallel existence of two irreversible reactions is of the greatest importance in metabolic control: it means that the direction of flux can be controlled by the activities of the two enzymes. A single reversible reaction could not be controlled in this way, because a catalyst cannot affect the direction of flux through a reaction, which is determined solely by thermodynamic considerations.

If both reactions were to proceed uncontrolled at similar rates, there would be no net interconversion of F6P and FDP, but continuous hydrolysis of ATP, resulting eventually in death. This situation is known as a *futile cycle*, and in order to prevent it it is necessary either to segregate the two processes into different cells, or to control both enzymes so that each is active only when the other is inhibited. Although control is achieved by compartmentalization to some extent, it is not possible in tissues such as kidney and liver that can carry out both glycolysis and gluconeogenesis and some degree of control over the activities of the two enzymes is therefore essential in these tissues.

We must now consider whether an enzyme that obeys the ordinary laws of enzyme kinetics can be controlled sufficiently precisely. For an enzyme that obeys the Michaelis–Menten equation, $v = Vs/(K_m + s)$, a simple calculation shows that an increase in velocity from $0.1V$ to $0.9V$ requires an increase in substrate concentration from $K_m/9$ to $9K_m$, i.e. an 81-fold increase, in order to bring about a comparatively modest increase in rate. Similarly, for any enzyme that obeys the equation for simple competitive inhibition, $v = Vs/[K_m(1+i/K_i)+s]$, an 81-fold increase in inhibitor concentration is required in order to reduce the velocity from 90% to 10% of the uninhibited value. These effects are amplified to some extent by altering inhibitor and substrate concentrations in concert; for example, with the above equation an increase in s from $K_m/3$ to $3K_m$ accompanied by a decrease in i from $3K_i$ to $K_i/3$ brings about a 9-fold increase in v from $V/13$ to $9V/13$. However, even if there are several effectors acting in concert, the qualitative situation is the same: a drastic change in the environment is necessary in order to bring about even a modest change in rate. The requirements of metabolism are exactly the opposite: on the one hand, the concentrations of major metabolites must be maintained within small tolerances, and on the other hand, reaction rates must be capable of changing very greatly—probably more than the $0.1V$ to $0.9V$ range we have considered in many cases—in response to fluctuations within these small tolerances.

Clearly, the ordinary laws of enzyme kinetics are inadequate for providing the degree of control that is necessary for metabolism. Instead, many of the enzymes at control points display the property of responding with exceptional sensitivity to changes in metabolite concentrations. This property is generally known as *co-operativity*, because it is thought to arise in many instances from 'co-operation' between the active sites of polymeric enzymes. This chapter deals principally with the examination of the main theories that have been proposed in order to account for co-operativity.

The F6P–FDP interconversion reactions illustrate another important aspect of metabolic control, namely the fact that the immediate and ultimate products of a reaction may be different. Although ATP is a substrate of the phosphofructokinase reaction, the effect of glycolysis as a whole is to generate ATP, in very large amounts if glycolysis is considered as the route into the tricarboxylic acid cycle and electron transport. Thus ATP must be regarded as a product of glycolysis, even though it is a substrate of the reaction at which glycolysis is controlled. Hence ordinary product inhibition of phosphofructokinase by ADP produces the opposite effect from what is required: in order to permit the steady supply of energy, phosphofructokinase ought to be inhibited by the ultimate product of the pathway, ATP, as in fact it is. This type of inhibition cannot be provided by the usual mechanisms, i.e. by binding the inhibitor as a structural analogue of a substrate: in some cases these would produce an unwanted effect, while in others the ultimate product of a pathway may bear very little structural resemblance to any of the reactants in the controlled step, e.g. L-histidine bears very little similarity to phosphoribosyl pyrophosphate, its biosynthetic precursor. In order to permit inhibition or activation by metabolically appropriate effectors, many controlled enzymes have developed sites for effector binding that are separate from the catalytic sites. These sites are called *allosteric* sites, from the Greek for 'another solid,' in order to emphasize the structural dissimilarity between substrate and effector, and enzymes that possess them are called allosteric enzymes.

Many allosteric enzymes are also co-operative, and *vice versa*, because both properties are important in metabolic control. However, this does not mean that the two terms are interchangeable: they describe two different properties and should be clearly distinguished. In many cases, the two properties have been recognized separately: haemoglobin was known to be co-operative for over 60 years before the allosteric effect of 1,2-diphosphoglycerate was recognized; the first enzyme in the biosynthesis of histidine has long been known to be allosteric, but it has not been reported to be co-operative.

7.2 Binding of oxygen to haemoglobin

Although haemoglobin is not an enzyme but a transport protein, it would be absurd to discuss co-operativity without first discussing haemoglobin. Firstly, its co-operative properties were recognized (Bohr, 1903) long before those of any enzyme, and a large part of the effort in developing theories to account for co-operativity has been directed specifically at understanding the co-operativity of haemoglobin. Secondly, the binding of oxygen to haemoglobin can be directly measured at equilibrium, so that one does not have to rely on any questionable assumptions about the relationship between equilibrium binding and steady-state binding. Thirdly, haemoglobin differs from most co-operative enzymes in being fairly stable and easy to prepare in large amounts; it is thus very convenient for experimentation, and has been thoroughly studied. Finally, haemoglobin exists alongside myoglobin, a non-co-operative analogue used for storing oxygen in muscle, permitting a direct comparison that is impossible in any other case.

The binding of a ligand X to a simple monomeric protein E can be written as

$$E + X \overset{K}{\rightleftharpoons} EX$$

where K is the association constant, and the concentration of the complex at equilibrium is given by

$$[EX] = K[E][X] \qquad (7.1)$$

Before proceeding further, it is necessary to draw attention to two major differences between the symbols used in this chapter and those elsewhere in

118

the book. Firstly, equilibrium studies, particularly for haemoglobin, are usually discussed in terms of association constants rather than the dissociation constants that are more familiar to biochemists. This simplifies the appearance of many equations and, in any case, conversion of results from the literature to a different system would probably create more confusion than it would avoid. However, one major theory of co-operativity, the symmetry model of Monod, Wyman and Changeux (Section 7.7), is always discussed in terms of dissociation constants. Secondly, in discussing co-operativity, abbreviated symbols for concentrations, such as x for $[X]$, will not be used because it is difficult to adapt this system to the concentrations of complicated species, such as EX_4, and because in equilibrium studies the total protein concentration is often of the same order of magnitude as the total ligand concentration. As a result, the free ligand concentration may be much smaller than the total ligand concentration, in contrast to the usual situation in steady-state kinetics.

We can define a quantity Y, known as the *fractional saturation*, as the fraction of binding sites that are occupied at any instant, i.e.

$$Y = \frac{\text{number of occupied binding sites}}{\text{total number of binding sites}} = \frac{[EX]}{[E]+[EX]}$$

Then, from equation 7.1, we obtain

$$Y = \frac{K[X]}{1+K[X]} \tag{7.2}$$

This equation is the Langmuir isotherm (cf. Section 2.2), and describes a rectangular hyperbola through the origin when Y is plotted against $[X]$, which approaches $Y = 1$ when $[X]$ is large. Thus it closely resembles the Michaelis–Menten equation, with $1/K$ replacing K_m, and 1 replacing V (by definition).

If the fractional saturation of myoglobin is measured as a function of the partial pressure of oxygen, the results do indeed obey equation 7.2, but the results for a similar experiment with haemoglobin fall on a different curve, which is strikingly sigmoid or S-shaped, as illustrated in *Figure 7.1*. Equation 7.2 cannot account for this curve and, from the time of Hill (1910) onwards, much effort has been devoted to the search for a plausible physical model that will account for it.

Before discussing the various models for co-operativity that have been proposed, it is instructive to consider an important physical difference between myoglobin and haemoglobin. Myoglobin is a monomer, with a single polypeptide chain and a single oxygen-binding site per molecule, but haemoglobin is a tetramer, consisting of four polypeptide chains, or *subunits*, per molecule, each with an oxygen-binding site. Moreover, although there are two distinct types of subunit in the haemoglobin molecule, two α-subunits and two β-subunits, they are similar in structure, not only to one another but also to myoglobin; to a first order of approximation, haemoglobin resembles a tetramer of myoglobin. It seems obvious with hindsight that the differences in binding properties between the two proteins are related

119

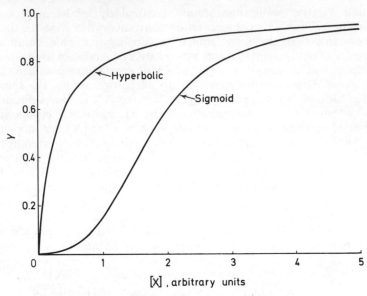

Figure 7.1 Comparison of a hyperbolic binding curve with a sigmoid curve

to their different degrees of association, but it is worth noting that the detailed information about the structures is rather recent (Kendrew *et al.*, 1960; Perutz *et al.*, 1960) and was not available to the earlier investigators of co-operativity.

7.3 Hill equation

Hill (1910) was concerned to account for the divergent values that he and a number of other investigators had observed for the molecular weight of haemoglobin. He suggested that monomeric haemoglobin might associate under the influence of salts, and that the oxygen-binding properties under different conditions might provide evidence of this association. If each polymer E_h binds h molecules of ligand in a single step:

$$E_h + hX \rightleftarrows E_h X_h$$

then the concentration of the complex $E_h X_h$ is given by

$$[E_h X_h] = K_h [E_h] [X]^h \qquad (7.3)$$

where K_h is the appropriate association constant. For a solution that contains a fraction λ of dimer and a fraction $(1 - \lambda)$ of monomer, the fractional saturation is given by

$$Y = \frac{\lambda K_2 [X]^2}{1 + K_2 [X]^2} + \frac{(1 - \lambda) K_1 [X]}{1 + K_1 [X]}$$

Hill found that this equation fitted some of the available binding data within experimental error, but he felt that it was unreasonable to restrict the model

120

to monomer and dimer and that a more realistic model should include higher polymers, but this led to an equation that was far too complicated to be manageable. Accordingly, he suggested, as a purely empirical equation, the following:

$$Y = \frac{K_h[X]^h}{1 + K_h[X]^h} \tag{7.4}$$

This is the equation that one would obtain from equation 7.3 by assuming that the protein exists solely in the two forms E_h and E_hX_h. Hill found that this equation, now known as the *Hill equation*, fitted all of the available data very accurately, with different values of h in the range 1.0–3.2.

If equation 7.4 is rearranged as follows:

$$\frac{Y}{1-Y} = K_h[X]^h$$

$$\log\left(\frac{Y}{1-Y}\right) = \log K_h + h \log [X] \tag{7.5}$$

it can be seen that a plot of $\log\left(\frac{Y}{1-Y}\right)$ against $\log[X]$ should be a straight line of slope h. This plot, which is illustrated in *Figure 7.2*, is known as the

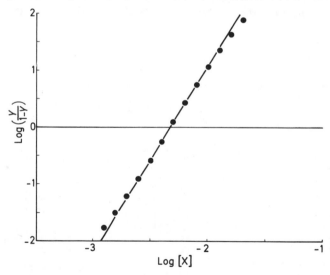

Figure 7.2 Hill plot: The line is drawn according to the Hill equation (equation 7.5), and does not fit the points exactly except in the middle of the range. It is often difficult to make measurements outside the range −1 to +1 on the ordinate, which corresponds to a range 0.09–0.91 in the value of Y

Hill plot and provides a simple means of evaluating h and K_h. It has been found to fit a wide variety of binding data remarkably well for values of Y in the range 0.1–0.9, but deviations always occur at the extremes (as indicated in *Figure 7.2*) because equation 7.4 is at best only an approximation of a more complex relationship.

121

Hill was careful to disclaim any physical meaning for K_h and h, but many more recent workers have supposed that h ought to be equal to n, the number of subunits in the fully associated protein. They have accordingly been puzzled that h is generally non-integral, and is rarely equal to n. However, as there is no reason why h should be an integer, it is in no way surprising if it is not. For reasons that will become clear when the Adair equation is examined in the next section, h cannot exceed n, so it does provide a lower limit for n.

The exponent h is now generally known as the *Hill coefficient*. It is widely used as an index of co-operativity, as the greater the value of h the greater is the degree of co-operativity. Taketa and Pogell (1965) have suggested a different parameter, the *co-operativity index*, R_x, defined as the ratio of the [X] values required for $Y = 0.9$ and $Y = 0.1$. Thus R_x has a more obvious experimental meaning than h, and is more convenient for discussing the properties of co-operative proteins in relation to their physiological role. R_x has the further advantage that it is a purely empirical measure, not based on any theoretical model of dubious validity. The relationship between the two indexes can be obtained by substituting $Y = 0.1$ and $Y = 0.9$ into equation 7.4 and solving for [X] in each case:

$$R_x = 81^{1/h}$$

This expression is only as accurate as equation 7.4, of course, but that is sufficient for most purposes. Some values calculated from it are shown in *Table 7.1.*

Table 7.1 RELATIONSHIP BETWEEN THE TWO INDEXES OF CO-OPERATIVITY

The table shows the relationship between the Hill coefficient, h, and the co-operativity index, R_x, suggested by Taketa and Pogell (1965). The values are calculated on the assumption that the Hill equation holds exactly.

h	R_x	Qualitative
0.5	6 560	Negatively co-operative
0.6	1 520	
0.7	533	
0.8	243	
0.9	132	
1.0	81.0	Non-co-operative
1.5	18.7	(Positively) co-operative
2.0	9.00	
2.5	5.80	
3.0	4.33	
3.5	3.51	
4.0	3.00	
5.0	2.41	
6.0	2.08	
8.0	1.73	
10	1.55	
15	1.34	
20	1.25	

7.4 Adair equation

Adair (1925*a,b*), after determining that the molecular weight of haemoglobin was about four times greater than had previously been thought, suggested that there were four oxygen-binding sites per molecule, and that these sites were filled in a four-step process, rather than in the concerted manner assumed by Hill. In this case we have:

$$E + X \underset{\longleftarrow}{\overset{4K_1}{\longrightarrow}} EX$$

$$EX + X \underset{\longleftarrow}{\overset{\frac{3}{2}K_2}{\longrightarrow}} EX_2$$

$$EX_2 + X \underset{\longleftarrow}{\overset{\frac{2}{3}K_3}{\longrightarrow}} EX_3$$

$$EX_3 + X \underset{\longleftarrow}{\overset{\frac{1}{4}K_4}{\longrightarrow}} EX_4$$

where the association constants are *intrinsic* (site) constants rather than molecular constants. When defined in this way, all four constants would be equal, i.e. $K_1 = K_2 = K_3 = K_4$, if the four binding sites were identical and acted independently of one another.

For Adair's model, the concentrations of the various species are given by:

$$\left.\begin{aligned}
[EX] &= 4K_1[E][X] \\
[EX_2] &= \tfrac{3}{2}K_2[EX][X] = 6K_1K_2[E][X]^2 \\
[EX_3] &= \tfrac{2}{3}K_3[EX_2][X] = 4K_1K_2K_3[E][X]^3 \\
[EX_4] &= \tfrac{1}{4}K_4[EX_3][X] = K_1K_2K_3K_4[E][X]^4
\end{aligned}\right\} \quad (7.6)$$

These lead simply to the following expression for the fractional saturation:

$$
\begin{aligned}
Y &= \frac{\text{number of occupied sites}}{\text{total number of sites}} = \frac{[EX]+2[EX_2]+3[EX_3]+4[EX_4]}{4([E]+[EX]+[EX_2]+[EX_3]+[EX_4])} \\
&= \frac{K_1[X]+3K_1K_2[X]^2+3K_1K_2K_3[X]^3+K_1K_2K_3K_4[X]^4}{1+4K_1[X]+6K_1K_2[X]^2+4K_1K_2K_3[X]^3+K_1K_2K_3K_4[X]^4}
\end{aligned}
$$

$$(7.7)$$

This result is known as the *Adair equation* for four sites; similar equations can be derived in an obvious way for other numbers of sites. If all four intrinsic association constants are equal, the Adair equation simplifies to

$$Y = \frac{K_1[X](1+K_1[X])^3}{(1+K_1[X])^4} = \frac{K_1[X]}{1+K_1[X]} \qquad (7.8)$$

i.e. equation 7.2, the equation of a hyperbola. However, as the binding curve for haemoglobin is not a hyperbola, all four association constants cannot be equal for haemoglobin.

If K_4 is sufficiently large compared with K_1, K_2 and K_3, equation 7.7 simplifies to

123

$$Y = \frac{K_1 K_2 K_3 K_4 [X]^4}{1 + K_1 K_2 K_3 K_4 [X]^4}$$

i.e. the Hill equation with $K_h = K_1 K_2 K_3 K_4$ and $h = 4$. However, there is no way in which equation 7.7 can be simplified to yield a Hill coefficient greater than 4. Moreover, if $[X]$ is made sufficiently small, then, whatever the values of the association constants, $K_1[X]$ must eventually become greater than the higher order terms in equation 7.7, and so it must simplify to

$$Y = \frac{K_1[X]}{1 + 4K_1[X]}$$

at very low $[X]$. This equation yields a Hill plot in which h approaches unity as $[X]$ approaches zero. Similarly, $h \to 1$ as $[X] \to \infty$. Thus, in general, for any values of the association constants, the Hill coefficient must approach unity at the extremes of $[X]$, and cannot exceed the number of binding sites for intermediate values of $[X]$.

Adair's model is the most general model possible for a ligand binding to a pure non-associating protein at equilibrium. In this context, pure is taken to mean that different isomeric forms of the protein are permissible provided that all are in equilibrium and that the isomerization does not include dissociation or association. Let us consider Adair's model (equations 7.6) with the inclusion of an inactive species E' in equilibrium with E according to the following equation:

$$E \overset{K_0}{\underset{\longleftarrow}{\longrightarrow}} E'; \qquad [E'] = K_0[E]$$

In this case, the fractional saturation is given by

$$
\begin{aligned}
Y &= \frac{[EX] + 2[EX_2] + 3[EX_3] + 4[EX_4]}{4([E] + [E'] + [EX] + [EX_2] + [EX_3] + [EX_4])} \\
&= \frac{K_1[X] + 3K_1 K_2[X]^2 + 3K_1 K_2 K_3[X]^3 + K_1 K_2 K_3 K_4[X]^4}{(1 + K_0) + 4K_1[X] + 6K_1 K_2[X]^2 + 4K_1 K_2 K_3[X]^3 + K_1 K_2 K_3 K_4[X]^4}
\end{aligned}
$$

$$(7.9)$$

If every term in this equation is divided by $1 + K_0$, it is seen to be identical in form with the Adair equation (equation 7.7) with the replacement of K_1 by $K_1/(1 + K_0)$. The results are similar if any or all of the other species EX, EX_2 etc. isomerize. Thus the form of the binding equation is unaffected by isomerization; conversely, the form of the binding equation cannot provide evidence of isomerization.

This result can be contrasted with the effect of a species E' that is not in equilibrium with E: in this case, the fractional saturation is given by

$$Y = \frac{K_1[X] + 3K_1 K_2[X]^2 + 3K_1 K_2 K_3[X]^3 + K_1 K_2 K_3 K_4[X]^4}{([E']/[E] + 1) + 4K_1[X] + 6K_1 K_2[X]^2 + 4K_1 K_2 K_3[X]^3 + K_1 K_2 K_3 K_4[X]^4}$$

This equation is not of the form of equation 7.7: division of all terms by $[E']/[E] + 1$ is not equivalent to division by $1 + K_0$ in equation 7.9, because $[E]$ is not a constant but varies with $[X]$.

124

In the more general case of a mixture of proteins, each capable of binding ligands with different binding constants, the binding equation is very complicated, with terms to $[X]^{\Sigma n}$, where Σn is the total number of different binding sites. As this equation is then of similar form to the Adair equation for a pure protein with Σn binding sites, it might seem that such a mixture could display greater co-operativity than any of the pure species that compose it. However, this is not so because, in the absence of interactions between binding sites, ligands bind most strongly to the sites with the highest affinity, as is intuitively obvious. Thus, although mixtures do display deviations from the binding equations for pure proteins, the deviations are always in the direction of more negative co-operativity, i.e. towards higher values of R_x or smaller values of h.

It follows from this discussion that the presence of square or higher order terms in a binding equation provides no guarantee of co-operativity. In order to demonstrate this more precisely, let us consider the Adair equation for a protein with two binding sites:

$$Y = \frac{K_1[X] + K_1 K_2 [X]^2}{1 + 2K_1[X] + K_1 K_2 [X]^2}$$

from which we find

$$\log\left(\frac{Y}{1-Y}\right) = \log K_1 + \log[X] + \log\left(\frac{1 + K_2[X]}{1 + K_1[X]}\right)$$

Differentiating with respect to $\log[X]$, we obtain

$$h = 1 + \frac{(K_2 - K_1)[X]}{(1 + K_1[X])(1 + K_2[X])}$$

From this equation, it follows that $h > 1$ if $K_2 > K_1$; $h = 1$ if $K_2 = K_1$; and $h < 1$ if $K_2 < K_1$. Each of these results applies for all values of $[X]$, although the numerical value of h does vary with $[X]$, and approaches unity for very high or very low values of $[X]$. The co-operative case $K_2 > K_1$ is impossible unless the sites interact, because if the tighter binding site exists in the absence of ligand, the first ligand molecule will bind to it rather than the weaker site. Hence the observation of co-operativity provides clear evidence of interactions between binding sites, but the observation of negative co-operativity does not provide evidence of negative interactions unless it is first established that the protein is pure, because negative co-operativity is to be expected for non-identical binding sites that do not interact. Similar conclusions apply to more complex binding equations.

7.5 Pauling's treatment

Adair's treatment of the oxygen–haemoglobin system fits the observed behaviour at least as well as any other model that has been proposed since, but it is unsatisfying, because it offers no physical explanation of the interactions that give rise to the co-operativity. Several explanations have been presented in recent years, but one earlier treatment ought first to be men-

125

tioned briefly, namely that of Pauling (1935). At a time when no information existed about the geometry of the haemoglobin molecule, Pauling postulated that the four haem groups (the sites of oxygen binding) interacted with one another in pairs, in such a way that the binding of oxygen to one member of a pair caused the association constant for binding at the other member to be increased by a factor that was the same for every pair. He found that if each haem interacted in this way with only one other haem, i.e. if the molecule contained two independent pairs, the resulting equation could not account for the observations, but if each interacted with two others, as the corners of a square, a close fit was possible if the interaction factor was properly chosen; similarly good results were possible if each haem interacted with three others, as the vertices of a tetrahedron. The mathematical treatment of Pauling's model is identical with that of the sequential model and it is therefore not discussed here but is deferred to Section 7.8.

The conceptual advantage of Pauling's model over Adair's, is that it requires only two constants, each of which has a clear physical meaning: one is the unperturbed association constant for each site and the other is the perturbation factor brought about by oxygen binding at the interacting binding site. Provided that the haems were assumed to be sufficiently close together to interact electronically, there was no difficulty in accounting for the interaction in pairs, as similar interactions were (and are) well known in simple systems, in fact so well known that Pauling felt no need to discuss their physical nature in any detail. Consider, for example, the 'binding' of electrons to quinone:

In principle, the free-radical intermediate should occur, but the electronic structure is such that the intermediate is greatly disfavoured at the expense of the two extremes; in effect, the binding is exceedingly co-operative, in accordance with the Hill equation for two sites. Negative co-operativity is even more common in simple systems, such as the binding of two protons to symmetrical dianions such as oxalate or succinate.

Pauling reasoned that if the haem groups were arranged tetrahedrally on the surface of the haemoglobin molecule they would be too far apart (about 4.7 nm) to interact, but if they were placed in a square on one side of the apoprotein they could be as close to one another as necessary. He therefore preferred the square arrangement, although he recognized that both arrangements fitted the experimental results equally well.

With the determination of the three-dimensional structure of haemoglobin (Perutz et al., 1960), it became clear that the haem groups were much too far apart (2.5–4.0 nm) to interact in the way that Pauling had envisaged, even if

126

each interacted with only one other. Moreover, there was no conjugated system of chemical bonds to permit long-range interactions (as found, for example, in vitamin A and many other coloured molecules). This discovery did not simply overthrow Pauling's explanation of co-operativity; it also overthrew all others, including Adair's, because all explanations included, either explicitly or implicitly, the concept of haem–haem interactions. The phenomenon of co-operativity thus became more mysterious than it had been since Hill's pioneering work in 1910, just when it was becoming clear that it was by no means confined to haemoglobin but was of the greatest importance in the regulation of metabolism.

7.6 Induced fit

All modern theories of co-operativity are derived from the *theory of induced fit* (Koshland, 1958, 1959*a,b*), in the sense that they all include the assumption that proteins are not rigid but can exist in a *limited* and *purposive* variety of conformations. It is appropriate, therefore, to digress slightly in order to consider the experimental and theoretical basis of induced fit.

The high degree of specificity that enzymes display towards their substrates has impressed biochemists since the earliest work, even before anything was known about the physical and chemical structures of enzymes. Fischer (1894) was particularly impressed with the ability of living organisms to discriminate totally between sugars that differed only slightly and at atoms remote from the site of reaction. In order to explain this ability, he proposed that the active site of an enzyme was a negative imprint of its substrate(s), and that it would catalyse the reactions only of compounds that fitted precisely. This is very similar to the mode of operation of an ordinary (non-Yale) key in a lock, and the theory has long been known as Fischer's *lock-and-key model* of enzyme action. For many years, it seemed to explain all of the known facts of enzyme specificity, but, as more detailed research was carried out there were numerous observations that were very difficult to account for in terms of a rigid active site of the type that Fischer had envisaged. For example, the common occurrence of enzymes for two-substrate reactions that require the substrates to bind in the correct order provides one type of evidence, as mentioned in Section 5.2. A more striking example was the failure of water to react in several enzyme-catalysed reactions where one would certainly expect it to react. For example, hexokinase catalyses the phosphorylation of glucose:

$$\text{glucose} + \text{ATP} \rightarrow \text{glucose 6-phosphate} + \text{ADP}$$

The enzyme is not particularly specific: not only glucose, but also fructose, mannose and other sugars react fairly rapidly. However, water does not react, yet it can scarcely fail to saturate the enzyme, at a concentration of 56 M, about 7×10^6 times the Michaelis constant for glucose, and chemically it is at least as reactive a compound as the sugars that do react.

Koshland argued that these and other observations provided strong evidence for a *flexible* active site; he proposed that the active site of an enzyme

127

has the potential to fit the substrate precisely, but that it does not adopt the negative substrate form until the substrate binds. This conformational change accompanying substrate binding brings about the proper alignment of the catalytic groups of the enzyme with the site of reaction in the substrate. With this hypothesis, the properties of hexokinase can easily be explained: water can certainly bind to the active site of the enzyme, but it lacks the bulk to bring about the conformational change necessary for catalysis.

Koshland's theory is known as the induced-fit hypothesis, to emphasize its difference from Fischer's theory, which assumes that the fit between enzyme and substrate pre-exists and does not need to be induced. The lock-and-key analogy can be pursued a little further by likening Koshland's conception to a Yale lock, in which the key must not merely fit but must also realign the tumblers before it will turn.

The induced-fit theory has had important consequences in several branches of enzymology, but it was particularly important in the understanding of allosteric and co-operative phenomena in proteins, because it provided a simple and plausible explanation of long-range interactions. Provided that a protein combines rigidity with flexibility in a purposive and controlled way, like a pair of scissors, a substrate-induced conformational change at one point in the molecule may be communicated over several nanometres to any other point.

7.7 Symmetry model of Monod, Wyman and Changeux

Both co-operative interactions in haemoglobin and allosteric effects in many enzymes require interactions between sites that are widely separated in space. A striking example of this requirement is provided by the allosteric inhibition of phosphoribosyl-ATP pyrophosphorylase by histidine: Martin (1963) found that mild treatment of the enzyme with mercury(II) ions destroyed the sensitivity of the catalytic activity to histidine but did not affect either the uninhibited activity or the binding of histidine. In other words, the metal ion interfered neither with the catalytic site nor with the allosteric site, but with the connection between them. Monod, Changeux and Jacob (1963) studied many examples of co-operative and allosteric phenomena, and concluded that they were closely related and that conformational flexibility probably accounted for both. Subsequently, Monod, Wyman and Changeux (1965) proposed a general model in order to explain both phenomena within a simple set of postulates. The model is often referred to as the *allosteric model*, but the term *symmetry model* is preferable because it emphasizes the principal difference between it and later models and because it avoids the contentious association between allosteric and co-operative phenomena.

The symmetry model starts from the observation that many, perhaps all, co-operative proteins contain several subunits in each molecule. (For simplicity we shall assume a tetrameric protein, but any number of subunits greater than one is possible.) The model includes the following postulates:

(1) Each subunit can exist in two different conformations, designated R (relaxed) and T (tense).

(2) All subunits of a molecule must occupy the same conformation at any time; hence, for a tetrameric protein, the conformational states R_4 and T_4 are the only two permitted, conformational mixtures such as R_3T being forbidden.

(3) The two states of the protein are in equilibrium, with an equilibrium constant $L = [T_4]/[R_4]$.

(4) A ligand can bind to a subunit in either conformation, but the dissociation constants are different: $K_R = [R][X]/[RX]$ for each R subunit, $K_T = [T][X]/[TX]$ for each T subunit, and $c = K_R/K_T$ by definition.

These postulates imply the following set of equilibria between the various forms of the protein:

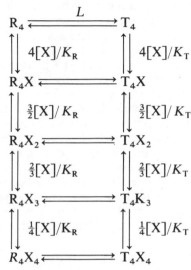

The concentrations of the ten species are related by the following expressions:

$$[R_4X] = 4[R_4][X]/K_R$$
$$[R_4X_2] = \tfrac{3}{2}[R_4X][X]/K_R = 6[R_4][X]^2/K_R^2$$
$$[R_4X_3] = \tfrac{2}{3}[R_4X_2][X]/K_R = 4[R_4][X]^3/K_R^3$$
$$[R_4X_4] = \tfrac{1}{4}[R_4X_3][X]/K_R = [R_4][X]^4/K_R^4$$
$$[T_4] = L[R_4]$$
$$[T_4X] = 4[T_4][X]/K_T = 4Lc[R_4][X]/K_R$$
$$[T_4X_2] = 6[T_4][X]^2/K_T^2 = 6Lc^2[R_4][X]^2/K_R^2$$
$$[T_4X_3] = 4[T_4][X]^3/K_T^3 = 4Lc^3[R_4][X]^3/K_R^3$$
$$[T_4X_4] = [T_4][X]^4/K_T^4 = Lc^4[R_4][X]^4/K_R^4$$

In each equation, the 'statistical' factor 4, $\tfrac{3}{2}$, etc., results from the fact that the dissociation constants are defined in terms of individual sites but the expressions are written for complete molecules. For example, $K_R = [R][X]/[RX]$

$= \frac{3}{2}[R_4X][X]/[R_4X_2]$, because there are three unliganded R subunits in each R_4X molecule and two liganded R subunits in each R_4X_2 molecule. The fractional saturation is now given by the equation

$$Y = \frac{\begin{array}{c}[R_4X]+2[R_4X_2]+3[R_4X_3]+4[R_4X_4]\\+[T_4X]+2[T_4X_2]+3[T_4X_3]+[T_4X_4]\end{array}}{\begin{array}{c}4([R_4]+[R_4X]+[R_4X_2]+[R_4X_3]+[R_4X_4]\\+[T_4]+[T_4X]+[T_4X_2]+[T_4X_3]+[T_4X_4])\end{array}}$$

$$= \frac{(1+[X]/K_R)^3[X]/K_R+Lc(1+c[X]/K_R)^3[X]/K_R}{(1+[X]/K_R)^4+L(1+c[X]/K_R)^4} \tag{7.10}$$

The form of the saturation function defined by this equation depends on the values of L and c. It is instructive to consider first some extreme values of these constants. If $L = 0$, i.e. the T form does not occur under any conditions, the equation simplifies to $Y = [X]/(K_R+[X])$, i.e. the Langmuir isotherm. Similarly, if $L \to \infty$, then $Y = [X]/(K_T+[X])$. It therefore follows that deviations from hyperbolic binding occur only if both conformational states of the protein occur in significant amounts. This is reasonable, because if there is only one form of the protein the model is the same as Adair's model with independent and identical binding sites (cf. equation 7.8).

Hyperbolic binding also arises if the ligand binds equally well to both R and T states, i.e. $c = 1$, as in this case it does not matter which state is present. Apart from these special cases, equation 7.10 predicts co-operative binding, although this may not be obvious unless we consider the special case when $c = 0$, i.e. X binds *only* to the R state. This is a natural application of induced fit, but it is not an essential characteristic of the symmetry model as proposed by Monod, Wyman and Changeux. If $c = 0$, equation 7.10 simplifies to

$$Y = \frac{(1+[X]/K_R)^3[X]/K_R}{L+(1+[X]/K_R)^4} \tag{7.11}$$

At very high values of $[X]$, when $[X]/K_R \gg L$, the term L in the denominator is negligible and the expression as a whole factorizes to the Langmuir isotherm; but at low values of $[X]$ the term L dominates the denominator, so that the saturation curve rises very slowly from the origin. In other words, the curve must be sigmoid if L is large compared with 1.

To see why the more general expression, equation 7.10, predicts co-operativity, we must examine its relationship to the Adair equation. If the terms $(1+[X]/K_R)^3$, etc., in equation 7.10 are multiplied out and rearranged, the equation assumes the form of the Adair equation for four sites (equation 7.7), with the four Adair association constants defined as

$$K_1 = \frac{1+Lc}{K_R}$$

$$K_2 = \frac{1+Lc^2}{(1+Lc)K_R}$$

$$K_3 = \frac{1+Lc^3}{(1+Lc^2)K_R}$$

$$K_4 = \frac{1+Lc^4}{(1+Lc^3)K_R}$$

If we now examine the ratio of any pair of constants, e.g. K_3/K_2, we find

$$\frac{K_3}{K_2} = \frac{(1+Lc^3)(1+Lc)}{(1+Lc^2)^2} = \frac{1+Lc(c^2+1)+L^2c^4}{1+\quad 2Lc^2 \quad +L^2c^4}$$

As $Lc(c^2+1) \geqslant 2Lc^2$ for all positive values of L and c, it follows that the right-hand fraction cannot be less than 1, i.e. $K_3 \geqslant K_2$. Similar results apply to all other pairs of constants, and also apply if the model is generalized to include more than four binding sites and more than two conformations. The symmetry model must therefore give rise to positive co-operativity and cannot give rise to negative co-operativity. Some representative binding curves are shown in *Figure 7.3*.

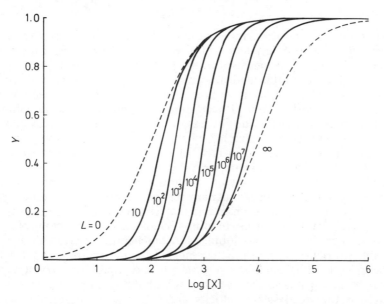

Figure 7.3 Binding curves for the symmetry model (equation 7.10), with c = 0.01 and L = 0–∞ as indicated: Arbitrary units are used for [X]. The curve for L = 0 is the binding curve for the pure R state, and the curve for L = ∞ is the binding curve for the pure T state. Both of these extreme curves would be hyperbolic (cf. Figure 7.1) if Y were plotted against [X] rather than log[X], whereas the intermediate curves would be sigmoid

Monod, Wyman and Changeux distinguished between *homotropic* effects, or interactions between identical ligands, and *heterotropic* effects, or interactions between different ligands, such as a substrate and an allosteric inhibitor. Although the model predicts that homotropic effects must be positively co-operative, no such restriction applies to heterotropic effects;

one of the most satisfying features of the symmetry model is the way in which it accounts for heterotropic effects with no extra complexity. Let us consider a case with three ligands, X, A and I, and let us assume that X and A bind only to the R conformation, with dissociation constants K_R and K_a, respectively, and that I binds only to the T conformation, with a dissociation constant K_i. (Note that for X this is the simplified case defined by equation 7.11. A more general treatment is possible but less instructive because its predictions are less obvious.) For this system, there are three saturation functions for the three ligands, which can be calculated in the same way as before. For the binding of X the result is

$$Y_x = \frac{(1+[X]/K_R)^3[X]/K_R}{L\left(\dfrac{1+[I]/K_i}{1+[A]/K_q}\right)^4 + (1+[X]/K_R)^4}$$

If we define $\Lambda = L(1+[I]/K_i)^4/(1+[A]/K_a)^4$, we can write this expression as

$$Y_x = \frac{(1+[X]/K_R)^3[X]/K_R}{\Lambda + (1+[X]/K_R)^4}$$

which is identical in form with equation 7.11 apart from the replacement of the constant L with Λ, a function of both $[I]$ and $[A]$. Clearly, I inhibits the binding of X, because Λ increases with $[I]$, whereas A assists the binding of X. Both allosteric inhibition and activation are therefore accounted for very neatly. However, in addition, as both I and A affect Λ, they also affect the degree of co-operativity and so, according to this model, allosteric inhibition should be accompanied by an increase in co-operativity, whereas allosteric activation should be accompanied by a decrease in co-operativity.

A number of enzymes do in fact behave in this way, as pointed out by Monod, Wyman and Changeux, provided that one can assume that steady-state binding is a true reflection of equilibrium binding. Unfortunately, there is no more justification for making any such assumption in a complex system than there is in a simple system, except that if it is not made experiments must be confined to true equilibrium binding systems such as the binding of oxygen and other ligands to haemoglobin. Most biochemists regard this as an intolerable restriction and in practice most of the information about heterotropic effects (and indeed homotropic effects in co-operative enzymes) comes from non-equilibrium studies of enzymes, with the assumption that $Y = v/V$, where v is the steady-state velocity and V is the asymptotic velocity at saturation.

The assumption that $Y = v/V$ is unfortunate, because it is difficult or impossible to prove, but it usually cannot be avoided because it is a hopeless task to try to interpret kinetic data for co-operative enzymes without making some simplifying assumptions. This is not, of course, a peculiarity of the symmetry model as all other models of co-operativity include some simplifying assumptions. The assumption that $Y = v/V$ has at least the advantage of being reasonably plausible, but some of the other assumptions of the symmetry model are not plausible. In particular, the central assumption of conformational symmetry is highly questionable, and frequent and confident

132

repetition has not made it less so; the model includes no molecular mechanism to account for it and there is no physical law that requires it. A second difficulty results from the need to treat many enzymes as *perfect K systems*, in the terminology of Monod, Wyman and Changeux; in other words, it is necessary to assume that, although the R and T states may differ in their affinity for substrate by a factor of 1000 or more, the rate constants for breakdown to products do not differ at all. The plausibility of the perfect K system has never been seriously discussed in the literature, and indeed it has often been presented as being so obvious that no discussion was required.

In spite of these strictures, the symmetry model was a major step forward in the understanding of protein co-operativity. Although there are now several examples in the literature of negative co-operativity, which the symmetry model cannot explain, there may well be some enzymes that agree closely with it. For example, Blangy, Buc and Monod (1968) used the symmetry model to provide a quantitative description of the co-operative properties of phosphofructokinase from *Escherichia coli*. This enzyme was particularly attractive for study, because it was possible to measure the binding of one substrate, fructose 6-phosphate, over a very wide range of concentrations of an allosteric activator, ADP, and an allosteric inhibitor, phosphoenolpyruvate. The detailed agreement between theory and experiment over the whole study represents a major success of the symmetry model.

7.8 Sequential model of Koshland, Némethy and Filmer

Although the symmetry model incorporates the idea of purposive conformational flexibility, it departs from the induced-fit theory in permitting ligands to bind to both R and T conformations, albeit with different binding constants. Koshland, Némethy and Filmer (1966) showed that a more orthodox application of induced fit could account for co-operativity equally well. Like Monod, Wyman and Changeux, they postulated two conformations, which they termed the A and B conformations (corresponding to the T and R conformations, respectively), but they assumed that the B conformation was induced by ligand binding so that X binds only to the B conformation and the B conformation exists only with X bound to it.

Koshland, Némethy and Filmer assumed that co-operativity arose because the properties of each subunit were modified by the conformational states of the neighbouring subunits. This assumption is implicit in the symmetry model, but it is emphasized in the sequential model, which is much more concerned with the details of interaction, and avoids the arbitrary assumption that all subunits must exist simultaneously in the same conformation. Hence conformational mixtures such as AB_3, A_2B_2, etc., are not merely allowed, but are required by the assumption of strict induced fit.

Because the symmetry model was not concerned with the details of subunit interactions, in the previous section there was no need to consider the geometry of subunit association, i.e. the quaternary structure of the protein. However, the sequential model does require the geometry to be considered, because different arrangements of subunits result in different binding equations. The emphasis on geometry and the need to treat each geometry separately have given rise to the widespread but erroneous notion that the sequential model is more general and more complicated than the symmetry model; but for any given geometry, the two models are about equally complex and neither is a special case of the other. Both models can be generalized into the same general model (Haber and Koshland, 1967), by relaxing the symmetry requirement of the symmetry model and the strict induced-fit requirement of the simplest sequential model, but it is questionable whether this is worthwhile, because the resulting equation is too complicated to use. In some contexts, it is helpful to refer to the ordinary form of the sequential model as the *simplest* sequential model, in order to distinguish it from the general model.

In discussing the sequential model, we shall consider the '*square*' geometry, in which each subunit is assumed to interact with its two neighbours, assuming that the four subunits are arranged in a square. Tetrahedral and linear arrangements are also possible for four subunits, but the method of analysis is the same as for the square case, although the results differ in detail. If the A conformation is shown as a circle and the B conformation as a square, the six possible species for the binding of X can be drawn as follows:

$$A_4 \qquad A_3BX \qquad A_2B_2X_2 \qquad AB_3X_3 \qquad B_4X_4$$

Note that there are two ways of drawing $A_2B_2X_2$, which must be considered separately, because the subunit contacts are different. The concentration of each species can be expressed by considering the various changes needed in order to obtain it from the standard state, A_4. For example, in order to obtain AB_3X_3 from A_4 the following changes must occur:

(1) Three subunits must undergo the conformational change $A \rightarrow B$. This is represented by K_t^3, where K_t is the notional equilibrium constant $[B]/[A]$ for an isolated subunit. In the simplest sequential model, K_t is tacitly assumed to be very small, in keeping with the assumption that the B conformation occurs only when induced by the binding of X.

(2) Three molecules of X must bind to three B subunits. This is represented by $K_x^3[X]^3$, where K_x is the association constant $[BX]/[B][X]$ for the binding of X to an isolated B subunit.

(3) In the square geometry there are four interfaces between neighbouring subunits. In the standard state, A_4, each interface can be designated AA, as both touching subunits are in the A conformation. However, in AB_3X_3 there are no AA interfaces; instead there are two AB interfaces and two BB interfaces and so K_{AB} represents the equilibrium constant $[AB]/[AA]$ for the conversion of an AA interface into an AB interface, and $K_{BB} = [BB]/[AA]$. Similarly, the net change requires $K_{AB}^2 K_{BB}^2$. The constant K_{AB} can also be regarded as an *absolute* measure of the

stability of the AB interface, but then another quantity, K_{AA}, is required for the AA interface, which is arbitrarily assigned a value of unity. It is simpler and just as rigorous to regard K_{AB} as a relative measure of the stability of the AB interface compared with the AA interface, and then no extra constant K_{AA} is needed. Similarly, it is simplest to regard K_{BB} as a measure of the stability of the BB interface compared with the AA interface.

(4) Finally, a statistical factor of 4 is required, because there are four equivalent ways of choosing three out of four subunits. The word *equivalent* is necessary here, because non-equivalent choices must be treated separately: for $A_2B_2X_2$ the statistical factor is 2 for the diagonal arrangement of ligands and 4 for the contiguous arrangement.

Multiplying all of these terms together, we obtain

$$[AB_3X_3] = 4[A_4]K_x^3K_t^3K_{AB}^2K_{BB}^2[X]^3$$

and similarly, for the other species

$$[A_3BX] = 4[A_4]K_xK_tK_{AB}^2[X]$$
$$[A_2B_2X_2] = 2[A_4]K_x^2K_t^2(2K_{AB}^2K_{BB} + K_{AB}^4)[X]^2$$
$$[B_4X_4] = [A_4]K_x^4K_t^4K_{BB}^4[X]^4$$

Combining all of these equations into an expression for the fractional saturation, we obtain

$$Y = \frac{[A_3BX] + 2[A_2B_2X_2] + 3[AB_3X_3] + 4[B_4X_4]}{4([A_4] + [A_3BX] + [A_2B_2X_2] + [AB_3X_3] + [B_4X_4])}$$

$$= \frac{K_xK_tK_{AB}[X] + K_x^2K_t^2(2K_{AB}^2K_{BB} + K_{AB}^4)[X]^2 + 3K_x^3K_t^3K_{AB}^2K_{BB}^2[X]^3 + K_x^4K_t^4K_{BB}^4[X]^4}{1 + 4K_xK_tK_{AB}^2[X] + 2K_x^2K_t^2(2K_{AB}^2K_{BB} + K_{AB}^4)[X]^2 + 4K_x^3K_t^3K_{AB}^2K_{BB}^2[X]^3 + K_x^4K_t^4K_{BB}^4[X]^4} \qquad (7.12)$$

As written, this equation is rather complicated because it allows for every aspect of ligand binding. However, it is less general than it appears, because some of the constants always occur in the same combinations; for example, $K_xK_t[X]$ always occurs as a product because of the assumption of strict induced fit. Less obvious combinations also occur because the subunit interactions are not independent of ligand binding; for example, K_{BB} cannot occur in a product that does not contain $K_x^2K_t^2[X]^2$, because a BB interaction implies the presence of two B subunits. Equation 7.12 can in fact be written in terms of only two constants:

$$Y = \frac{c^2\bar{K}[X] + c^2(2 + c^2)\bar{K}^2[X]^2 + 3c^2\bar{K}^3[X]^3 + \bar{K}^4[X]^4}{1 + 4c^2\bar{K}[X] + 2c^2(2 + c^2)\bar{K}^2[X]^2 + 4c^2\bar{K}^3[X]^3 + \bar{K}^4[X]^4} \qquad (7.13)$$

where $\bar{K}^4 = K_x^4K_t^4K_{BB}^4 - [B_4X_4]/[A_4][X]^4$ is the association constant for the complete four-step binding, and $c^2 = K_{AB}^2/K_{BB}$ is a measure of the stability of the AB interaction compared with the AA and BB interactions. It is useful to write the equation in this way, because it is the form required

135

for fitting it to experimental results; no fitting process is possible with the general form of the equation because, for example, any change in K_x can be exactly compensated for by an opposite change in K_t or K_{BB}. Some curves calculated from equation 7.13 are shown in *Figure 7.4.*

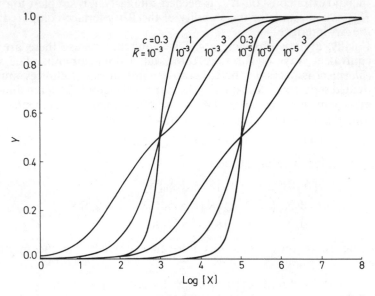

Figure 7.4 Binding curves for the sequential model (equation 7.13), with the values of $\overline{K}$ and c indicated: Arbitrary units are used for $[X]$. The location of the half-saturation point of each curve is determined solely by $\overline{K}$ and the shape is determined solely by c

The simpler equation is also useful for examining the relationship between the sequential model and the Adair equation. The four Adair constants are

$$K_1 = c^2\overline{K}$$
$$K_2 = \tfrac{1}{3}(2+c^2)\overline{K}$$
$$K_3 = 3\overline{K}/(2+c^2)$$
$$K_4 = \overline{K}/c^2$$

and if we consider the ratio of any pair of them, e.g. K_3/K_2, we find that it depends only on the value of c:

$$K_3/K_2 = 9/(2+c^2)^2$$

Therefore, $K_3 < K_2$ if $c > 1$ and $K_3 > K_2$ if $c < 1$, and similarly for other ratios of constants. As there is nothing in the definition of c that requires either of these conditions to hold rather than the other, the sequential model can account for negative co-operativity ($K_1 > K_2 > K_3 > K_4$) just as easily as positive co-operativity ($K_1 < K_2 < K_3 < K_4$), unlike the symmetry model, which is restricted to positive co-operativity.

Koshland, Némethy and Filmer showed that equation 7.13, and the corresponding equation for the tetrahedral geometry, fitted the oxygen–

136

haemoglobin saturation curve about as well as the equation for the symmetry model. It would therefore be difficult to distinguish between the models on the basis of the data for haemoglobin, or any other positively co-operative protein. However, this does not mean that saturation curves can never distinguish between models: for a protein showing negative co-operativity, the saturation curve alone provides sufficient evidence for ruling out the symmetry model. In fact, several examples of negative co-operativity have been reported in recent years and, although some of these results may reflect impure samples of protein, the binding of NAD to glyceraldehyde 3-phosphate dehydrogenase from rabbit muscle displays such an extreme degree of negative co-operativity (Conway and Koshland, 1968) that it cannot possibly be explained by impurities.

Although the sequential model was originally proposed as a way of accounting for homotropic interactions, Kirtley and Koshland (1967) subsequently extended it in order to account for heterotropic interactions also. Their treatment is not easy to summarize briefly, because they considered several possibilities, all of them plausible enough for one to expect each to occur with some proteins. However, their proposals are simple to understand and to apply to any given set of assumptions. They recognized that a second ligand might induce new conformations different from both A and B, but they considered that this would give rise to such complexity that a two-conformation model was preferable unless it could be shown to be inadequate. They therefore restricted their treatment to the same two conformations, A and B, and assumed that the main ligand of interest, X, could bind to the B conformation only. We shall consider only one of several possible modes of binding of a second ligand L, but it will be clear that other modes are conceivable and that they can be treated in the same way. Let us assume that the protein is a dimer, that L can bind to the B conformation only and that it competes for the same binding site as X, i.e. that X and L cannot bind simultaneously to the same subunit. With these assumptions, the possible species are as follows:

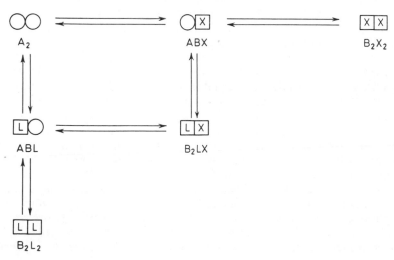

137

If we define K_x, K_t, K_{AB} and K_{BB} in the same way as before, and another association constant $K_L = [BL]/[B][L]$ for the binding of L to the B conformation, we can calculate the concentrations of all species as before. For example, the concentration of B_2LX is given by

$$[B_2LX] = 2[A_2]K_xK_LK_t^2K_{BB}[X][L]$$

We can combine all of the expressions into one for the fractional saturation of X:

$$Y_x = \frac{[ABX]+2[B_2X_2]+[B_2LX]}{2([A_2]+[ABX]+[B_2X_2]+[ABL]+[B_2LX]+[B_2L_2])}$$

$$= \frac{K_xK_tK_{AB}[X]+K_x^2K_t^2K_{BB}[X]^2+K_xK_LK_t^2K_{BB}[X][L]}{1+2K_xK_tK_{AB}[X]+K_x^2K_t^2K_{BB}[X]^2+2K_LK_tK_{AB}[L]}$$
$$2K_xK_LK_t^2K_{BB}[X][L]+K_L^2K_t^2K_{BB}[L]^2$$

At any particular value of $[L]$, this is of the form of the Adair equation for the binding of X, with association constants that vary with $[L]$:

$$K_1 = \frac{K_xK_t(K_{AB}+K_LK_tK_{BB}[L])}{1+2K_LK_tK_{AB}[L]+K_L^2K_t^2K_{BB}[L]^2}$$

$$K_2 = \frac{K_xK_tK_{BB}}{K_{AB}+K_LK_tK_{BB}[L]}$$

The ratio K_2/K_1 is

$$\frac{K_2}{K_1} = \frac{1+2K_LK_tK_{AB}[L]+K_L^2K_t^2K_{BB}[L]^2}{c^2+2K_LK_tK_{AB}[L]+K_L^2K_t^2K_{BB}[L]^2} \tag{7.14}$$

where c^2 is defined, as before, as K_{AB}^2/K_{BB}. It is clear from these equations that although L resembles an ordinary competitive inhibitor from the mechanistic point of view, its effect on the binding of X is complicated by the inclusion of subunit interactions. So far as the second site is concerned, L acts only as an inhibitor, as K_2 decreases monotonically as $[L]$ increases, but if $K_{BB} > 2K_{AB}^2$, i.e. if $c^2 < \frac{1}{2}$, L acts as an activator at low concentrations because then the numerator of the expression for K_1 initially increases with $[L]$ more steeply than the denominator. The effect of L on the co-operativity is fairly straightforward, as can be seen from equation 7.14: whether the binding in absence of L is positively or negatively co-operative, it tends monotonically towards non-co-operative as $[L]$ increases.

As there are many other equally plausible assumptions that one could make about the binding of L (e.g. it could bind only to the A conformation, or it could bind simultaneously with X, or it could induce a third conformation, or it could bind only when X was already bound), it is clear that a general treatment of the two-ligand case of the sequential model is much more complicated than the treatment for the symmetry model. However, it is questionable whether the symmetry model can truly be regarded as the simpler model, because its algebraic simplicity is achieved at the expense of conceptual simplicity: the assumption of conformational symmetry, elegant though it be, is essentially arbitrary. Moreover, the sequential model can be applied to kinetic experiments without the need for the other arbitrary feature of the symmetry model, namely the concept of the 'perfect K system.' In the sequential model, the substrate is assumed to bind only to the B conformation; hence all catalytic reactions occur in the same local environment, and it is reasonable for them to obey a single rate constant.

Finally, some mention should be made of the *concerted model*. This is not, in fact, another model, but is the name used by Koshland, Némethy and Filmer to the special case of the symmetry model defined by equation 7.11, which they discussed with their own symbolism in order to facilitate comparison with the sequential model. However, it is not a special case of the sequential model, and it is wrong to regard it as such.

7.9 Half-of-the-sites reactivity

Since negative co-operativity was first observed, numerous examples of it have been recognized (*see* Levitzki and Koshland, 1969). Of these, several have proved to be examples of '*half-of-the-sites reactivity*' (Levitzki, Stallcup and Koshland, 1971), an extreme type of negative co-operativity in which only half of the apparently identical subunits in a protein display reactivity towards substrate or other reagents. This is consistent with the sequential model, particularly if one takes account of the fact that the individual subunits of any protein are asymmetric, and so the contacts between subunits are unlikely to be all alike (Cornish-Bowden and Koshland, 1970). However, MacQuarrie and Bernhard (1971) have suggested a different explanation of half-of-the-sites reactivity involving '*pre-existing asymmetry*.' They argue that in some proteins the identical subunits may be assembled in such a way that they occupy two different types of environment, so that the protein molecule is asymmetric even in the absence of ligand. This is not in itself antithetical to the sequential approach, and Cornish-Bowden and Koshland (1971) have suggested, for different reasons based on a computer study of subunit association, that there is no *a priori* need for identical subunits to be placed in identical environments. Exponents of the sequential model (e.g. Stallcup and Koshland, 1973), however, have generally preferred, in the interests of simplicity, to assume a symmetrical arrangement of subunits unless there is a clear reason for rejecting it. The evidence in favour of pre-existing asymmetry has recently been discussed in detail (Seydoux, Malhotra and Bernhard, 1974) and will not be pursued here.

7.10 Other equilibrium models of co-operativity

Frieden (1967) and Nichol, Jackson and Winzor (1967) have independently suggested that co-operativity may in some cases result from the existence

139

of an equilibrium between protein forms in different states of aggregation, such as a monomer A and a tetramer B_4:

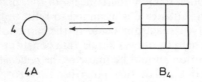

$4A$ B_4

If the two forms have different affinities for ligand, the model is conceptually rather similar to the symmetry model, and it predicts co-operative binding for much the same reasons. However, the equations that describe it are more complicated, because the protein concentration cannot be eliminated as it determines the relative amounts of A and B_4 in the unliganded state.

The association–dissociation model has the major advantage that it is far more accessible to experimental proof than those considered previously. If it applies, a large change in molecular weight should accompany ligand binding and the degree of co-operativity should depend on the protein concentration. Either of these results should be readily detectable, and neither is expected from the other models. Reversible association does appear to provide a complete explanation of the co-operativity observed for the binding of various nucleotides to glutamate dehydrogenase (Frieden and Colman, 1967). It may also contribute to the co-operativity of haemoglobin, which dissociates to a dimeric form at high salt concentrations, but it cannot provide a complete explanation of haemoglobin co-operativity, which is observed under many conditions where there is no dissociation. For any protein, it is an obvious precaution to check whether the observed co-operativity varies with protein concentration; if it does, then an association–dissociation model must be considered as a possible explanation.

7.11 Kinetic models of co-operativity

All of the models discussed have been essentially equilibrium models that can be applied to kinetic experiments only if v/V can be interpreted as a true measure of Y. However, there are also several purely kinetic models with no equilibrium counterparts. Of these models, the simplest is the following, proposed by Rabin (1967):

$$E + S \rightleftarrows ES$$
$$E' + S \rightleftarrows E'S \rightarrow E' + P$$

If the equilibrium between E and E' favours E, but the equilibrium between ES and E'S favours E'S (as indicated by the light and heavy arrows in the diagram), then S must bind more strongly to E' than to E because the second law of thermodynamics requires that the equilibrium constants for the

140

complete reaction $E + S \rightleftarrows E'S$ must be the same for both routes. Provided that the breakdown reaction $E'S \rightarrow E' + P$ is fast compared with the isomerizations $E' \rightarrow E$ and $ES \rightarrow E'S$, the binding of S cannot reach equilibrium. At low concentrations of S, E' decomposes before S can bind to it, but at high concentrations S can bind to E' before it has time to decompose. Thus the apparent affinity of the enzyme for substrate increases as the substrate concentration increases or, in other words, the binding is co-operative. It is important to the argument that the reaction $E'S \rightarrow E' + P$ must be fast, because if it is slow enough to be rate determining it cannot unbalance the substrate binding equilibrium, which cannot then be co-operative.

As has been seen in Section 5.2, the full steady-state rate equation for a two-substrate reaction proceeding through a ternary complex contains terms in the squares of the substrate concentrations if the substrates can bind to the enzyme in random order. Ferdinand (1966) has suggested that these square terms can give rise to a sigmoid dependence of the rate on either substrate concentration, with suitable values of the rate constants. In principle, this possibility can be distinguished from equilibrium models of co-operativity by the absence of co-operativity in a true binding experiment. In practice, however, difficulties are likely because, although binding experiments are much easier to interpret than kinetic experiments, kinetic experiments are much easier to carry out accurately than binding experiments.

Deviations from the Michaelis–Menten equation ought to occur with many other mechanisms that contain branched pathways (Sweeny and Fisher, 1968), but in practice they may be too slight to detect, as Gulbinsky and Cleland (1968) found with galactokinase (cf. Sections 3.7 and 5.2). At present, no enzyme is known to be co-operative for this type of reason, and it is likely that enzymologists will continue to use equilibrium models as long as it remains reasonable to do so.

Although kinetic models of the co-operativity of single enzymes may be of minor importance, kinetic control of metabolic pathways by means of two or more enzymes may be of the greatest importance in living organisms. Newsholme and Gevers (1967) have shown that a small amount of cycling at an important control point, such as the pair of opposing reactions catalysed by phosphofructokinase and fructose diphosphatase, can be used in order to provide much more sensitive control of the net flux through the pathway than the co-operative properties of a single enzyme can provide. This proposal also accounts for the presence of fructose diphosphatase in muscle, which was previously rather puzzling as muscle does not carry out gluconeogenesis. A cycle such as this is actually anything but 'futile,' as the 'wasted' ATP is put to a very valuable purpose.

8
Analysis of Progress Curves

8.1 Integrated rate equations

A kinetic equation can be written in two distinct forms, either as an expression that shows how the concentration of a reactant changes with time, or as one that shows how the rate of reaction varies with the concentrations of the reactants. The enzymologist normally considers the second of these to be the usual form, whereas to the chemist the first is the usual form. Chemists have, on the whole, continued to prefer integrated rate equations, which have the merit that they express what is actually measured. As has been seen in Chapter 2, however, the early workers in enzyme kinetics encountered many difficulties because they attempted to follow the usual chemical practice of fitting their observations to integrated equations. These difficulties were largely resolved when Michaelis and Menten (1913) showed that the behaviour of enzymes could be studied much more simply by measuring initial rates, when the complicating effects of product accumulation and substrate depletion did not apply. An unfortunate by-product of this early history, however, has been that biochemists have been very reluctant to use integrated rate equations, even when they have been appropriate.

In the early stages of an enzyme kinetic study, when the main problem is to discover what equation is obeyed and what the approximate values of the kinetic parameters are, the usual initial rate studies are appropriate. It is useful to be able to exclude product effects, or to introduce them in a controlled way. However, once these preliminaries have been completed, there are important advantages in following the reaction over an extended period and observing the progressive effects of accumulation of product and depletion of substrate. Much more information is contained in a progress curve than merely the extrapolated rate at zero time. It is possible, in principle, to obtain accurate values of the kinetic parameters from a fairly small number of experiments. There is also a major advantage in being able to avoid the subjective nature of estimating initial rates from a curved plot: it is almost impossible to do this reproducibly and without bias.

8.2 Integrated Michaelis–Menten equation

Before the Michaelis–Menten equation (or any other rate equation) can be integrated, it is necessary to introduce into it any relationships that exist between the reactant concentrations. Thus the substrate concentration, s, and the product concentration, p, are not independent, as their sum is a constant. Either s or p must be eliminated by means of the equation $s + p = s_0$. So the appropriate form of the Michaelis–Menten equation is equation 2.11, i.e.

$$\frac{dp}{dt} = \frac{V(s_0 - p)}{K_m + s_0 - p} \tag{8.1}$$

and therefore

$$\int \frac{(K_m + s_0)dp}{s_0 - p} - \int \frac{p\,dp}{s_0 - p} = \int V\,dt$$

giving

$$-(K_m + s_0)\ln(s_0 - p) + p + s_0\,\ln(s_0 - p) = Vt + \alpha$$

The condition $p = 0$ when $t = 0$ gives $\alpha = -K_m \ln s_0$, and so

$$Vt = p + K_m \ln[s_0/(s_0 - p)]$$

or

$$Vt = s_0 - s + K_m \ln(s_0/s) \tag{8.2}$$

Equation 8.2 is the integrated form of the Michaelis–Menten equation, and it defines the time course, or progress curve, of a reaction in which the Michaelis–Menten equation continues to be obeyed for an extended period after the start of the reaction. Instead of defining p in terms of t, as we might have wished, it defines t in terms of p, but this need not be a great inconvenience in practice. It is sensible to use the $\ln(s_0/s)$ form of this equation, and of other integrated rate equations, because this avoids any confusion when product is added at the start of the reaction, i.e. when $p_0 \neq 0$.

Like its differential form, equation 8.2 can be transformed in various ways in order to permit plotting as a straight line:

$$\frac{t}{\ln(s_0/s)} = \frac{1}{V}\left[\frac{s_0 - s}{\ln(s_0/s)}\right] + \frac{K_m}{V} \tag{8.3}$$

$$\frac{s_0 - s}{t} = V - \frac{K_m}{t}\ln(s_0/s) \tag{8.4}$$

$$\frac{t}{s_0 - s} = \frac{K_m}{V}\left[\frac{1}{(s_0 - s)}\right]\ln(s_0/s) + \frac{1}{V} \tag{8.5}$$

From equation 8.3, it can be seen that a plot of $t/\ln(s_0/s)$ against $(s_0 - s)/\ln(s_0/s)$ is a straight line with slope $1/V$ and intercept K_m/V on the $t/\ln(s_0/s)$ axis. Hence it resembles the plot of s/v against s in the initial rate case. Similarly, linear plots may be obtained from equations 8.4 and 8.5, which resemble the plots of v against v/s and of $1/v$ against $1/s$, respectively. In all three cases,

143

it is possible in principle to determine V and K_m from a single run. In order to do this effectively, it is necessary for s_0 to be well above K_m, and to continue to follow the reaction until $(s_0 - s)$ is well above K_m. For this method to be valid, it is also necessary to ensure that there is no significant product inhibition. In initial rate studies, product inhibition can usually be ignored if the reaction is followed only for a short time, when the product concentration can be made to approximate zero. On the other hand, if the reaction is to be followed over an extended period, product inhibition cannot simply be ignored. Moreover, *it is impossible to distinguish in a single run between the slowing down that results from depletion of substrate and any slowing down that results from competitive inhibition by the accumulating product.* The reason for this perhaps surprising fact will become clear in the next section. It should never be forgotten whenever equation 8.2 or its linear forms are used.

Of the three linear forms of equation 8.2, equation 8.4 seems to have been used most often, after it was first applied by Walker and Schmidt (1944) to studies of histidine ammonia-lyase. However, in order to be consistent with the rest of this book, we shall instead use equation 8.3 and similar equations. It is in fact slightly simpler than equation 8.4 to modify to the very important competitive product inhibition case.

8.3 Competitive product inhibition

Unless it is known that product inhibition is insignificant, it is safer to assume that it does occur and to calculate the product inhibition constant as part of the analytical strategy. Hence the following equation should be used, rather than equation 8.1, as the basic differential equation for progress curve studies:

$$\frac{dp}{dt} = \frac{V(s_0 - p)}{K_m(1 + p/K_p) + s_0 - p} \tag{8.6}$$

This equation is modified from equation 2.22, but a has been replaced with s in order to emphasize that we are exclusively concerned here with irreversible reactions; V^f, K_m^A and K_s^P are written simply as V, K_m and K_p for the same reason and s is written as $s_0 - p$ because s and p must be treated as variables and are related by the conservation equation $s + p = s_0$. If equation 8.6 is written in a slightly different form:

$$\frac{dp}{dt} = \frac{V(s_0 - p)}{K_m + s_0 - (1 - K_m/K_p)p} \tag{8.7}$$

it can be seen that it is of exactly the same form as equation 8.1 and it is for this reason that competitive product inhibition has an effect indistinguishable in a single run from that of substrate depletion. If several runs are carried out with different values of s_0, the cases become distinguishable, because the 'constant' $(K_m + s_0)$ varies with s_0. A slight qualification must be made to this point: the coefficient of p in the denominator of equation 8.1 is -1, but in equation 8.7 it is $-(1 - K_m/K_p)$, which can be zero if $K_m = K_p$, or positive if $K_m > K_p$. The case $K_m = K_p$ might seem to be such an unlikely event that

144

it could safely be ignored, but this is not so: if K_m is a good approximation to K_s, and if S and P possess closely similar groups that bind to the enzyme, it is possible that $K_p \approx K_m$. In this event, equation 8.7 takes the form

$$\frac{dp}{dt} = k(s_0 - p) = ks$$

where k represents $V/(K_m + s_0)$ and is a simple first-order rate constant (cf. Chapter 1). This case is, however, distinguishable from an ordinary first-order reaction by the fact that the rate 'constant' is a function of s_0.

Because equation 8.7 is of the same form as equation 8.1, it can be integrated in exactly the same way (Henri, 1903; Huang and Niemann, 1951; Schønheyder, 1952), and gives

$$Vt = (1 - K_m/K_p)(s_0 - s) + K_m(1 + s_0/K_p)\ln(s_0/s) \tag{8.8}$$

This equation can also be written in linear form as

$$\frac{t}{\ln(s_0/s)} = \frac{1}{V}(1 - K_m/K_p)\left(\frac{s_0 - s}{\ln(s_0/s)}\right) + \frac{K_m}{V}(1 + s_0/K_p) \tag{8.9}$$

and in two other ways, as before. In this case, a plot of $t/\ln(s_0/s)$ against $(s_0 - s)/\ln(s_0/s)$ again generates a straight line, but with slope $\frac{1}{V}(1 - K_m/K_p)$ and intercept $\frac{K_m}{V}(1 + s_0/K_p)$. Clearly, three constants cannot be determined from a single straight line, but they can readily be found from a series of experiments with different s_0 values. One method is to plot the intercepts of primary plots against s_0, which gives a straight line of slope K_m/VK_p and intercept K_m/V. Although the three kinetic constants may be calculated from these expressions, the relationships are not very convenient, and the following method, one of three described by Jennings and Niemann (1955), is much more elegant.

It should first be noted that the units of $t/\ln(s_0/s)$ and $(s_0 - s)/\ln(s_0/s)$ are the same as those of s_0/v and s_0, respectively. Consider now an extrapolation of the plot to $(s_0 - s)/\ln(s_0/s) = s_0$. Substitution into equation 8.9 gives

$$\frac{t}{\ln(s_0/s)} = \frac{1}{V}(1 - K_m/K_p)s_0 + \frac{K_m}{V}(1 + s_0/K_p)$$

$$= (s_0 + K_m)/V = s_0/v_0$$

where v_0 is the initial velocity. Thus a simple extrapolation generates the point $(s_0, s_0/v)$, exactly as is determined and plotted in initial-rate studies. By plotting the results of several runs on a single plot, extrapolating each line as described, a plot of s_0/v_0 against s_0 results, from which V and K_m can be determined in the usual way. This type of plot is illustrated in *Figure 8.1*.

It might seem that this is an unnecessarily laborious way of generating an ordinary plot of s_0/v_0 against s_0, but in fact it has much to recommend it: not only does it provide information about product inhibition, but it also provides more accurate values of s_0/v_0 than are usually available from

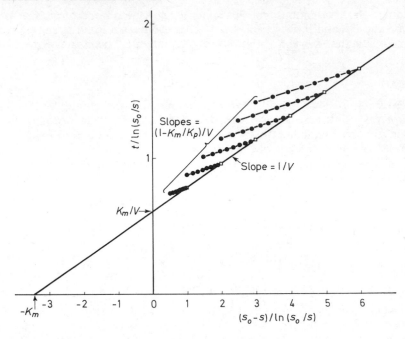

Figure 8.1 Determination of kinetic parameters from a series of six progress curves at different values of the initial substrate concentration, s_0, by plotting $t/\ln(s_0/s)$ against $(s_0 - s)/\ln(s_0/s)$: The enzyme was assumed to be subject to competitive product inhibition, and the experimental points (closed circles) were calculated from equation 8.9 after substituting the values $K_m = 3.46$, $K_p = 6.55$, $V = 5.7$ and $s_0 = 1-6$ (arbitrary units in each case). Points are shown for values of s from $0.9s_0$ to $0.2s_0$ at intervals of $0.1s_0$, i.e. from 10% to 80% completion of the reaction. For each value of s_0, the open square is obtained by extrapolating the line through the experimental points back to $(s_0 - s)/\ln(s_0/s) = s_0$, i.e. to 0% of reaction. These extrapolated points lie on a straight line of slope $1/V$ and intercepts $-K_m$ and K_m/V on the abscissa and ordinate, respectively (cf Figure 2.4)

initial-rate studies. The extrapolation required is very short, and can be carried out much more precisely and less subjectively than an estimate of the slope of a curve extrapolated back to zero time.

8.4 Inhibition by several products

In the previous section, competitive inhibition by a single product was considered. In practice, however, there may be several products, and the observed competitive inhibition constant is not the inhibition constant of any one product but the reciprocal of the sum of the reciprocals of all of the product inhibition constants. Hence in the previous section K_p should be replaced throughout with $1/(1/K_p + 1/K_q + \cdots)$. This follows from equation 8.6 because in this equation p/K_p should be replaced with $(p/K_p + q/K_q + \cdots)$, which is the same as $p(1/K_p + 1/K_q + \cdots)$ if all of the products are absent at the start of the reaction, so that $p = q = \cdots$ at all times.

It is desirable to be able to isolate the effects of the different products in a

146

reaction, which can be achieved by adding them separately in different amounts at the start of the reaction. Consider a case where there are two products, P and Q, of which only Q is added, with concentration q_0. Then,

$$\frac{dp}{dt} = \frac{V(s_0 - p)}{K_m\left(1 + \dfrac{p}{K_p} + \dfrac{p + q_0}{K_q}\right) + s_0 - p} \tag{8.10}$$

This equation can be integrated by the same method as before to give

$$Vt = K_m\left[1 + \frac{q_0}{K_q} + s_0\left(\frac{1}{K_p} + \frac{1}{K_q}\right)\right]\ln(s_0/s) + \left[1 - K_m\left(\frac{1}{K_p} + \frac{1}{K_q}\right)\right](s_0 - s)$$

And, as before, plots of $t/\ln(s_0/s)$ against $(s_0 - s)/\ln(s_0/s)$ can be constructed and extrapolated so as to generate a plot of s_0/v_0 against s_0. In this case, v_0 is the initial rate under the specified conditions, i.e. for Q present as a competitive inhibitor at a concentration q_0:

$$v_0 = \frac{Vs_0}{K_m(1 + q_0/K_q) + s_0} \tag{8.11}$$

and the value of K_q can be determined from the plot as described in Chapter 4. It is not necessary for the added inhibitor to be a product of the reaction for this procedure to be valid. If a competitive inhibitor, I, that is not a product is added at the start of the reaction at a concentration i, the inhibition constant, K_i, can be determined as follows. If i/K_i is substituted for $(p + q_0)/K_q$ in equation 8.10, the equation corresponding to equation 8.11 has i/K_i in place of q_0/K_q, and so K_i can be determined from the extrapolated points as described in Chapter 4.

8.5 Mixed inhibition by products

It was seen in Chapter 5 that product inhibition is not necessarily, or even usually, competitive. A more general case that ought to be considered is that where a product is a mixed inhibitor with competitive and uncompetitive inhibition constants K_p and K'_p, respectively:

$$\frac{dp}{dt} = \frac{V(s_0 - p)}{K_m(1 + p/K_p) + (s_0 - p)(1 + p/K'_p)}$$

This equation differs in form from the others that we have considered in this chapter, because of the p^2 term in the denominator, but it is still fairly easy to integrate:

$$\int \frac{(K_m + s_0)dp}{s_0 - p} + \int \frac{\left(\dfrac{K_m}{K_p} + \dfrac{s_0}{K'_p} - 1\right)p\, dp}{s_0 - p} - \int \frac{p^2\, dp}{K'_p(s_0 - p)} = \int V\, dt$$

147

Therefore,

$$-(K_m+s_0)\ln(s_0-p) - \left(\frac{K_m}{K_p} + \frac{s_0}{K_p'} - 1\right)[p+s_0\ln(s_0-p)]$$
$$-\frac{1}{K_p'}\left[\tfrac{1}{2}(s_0-p)^2 - 2s_0(s_0-p) + s_0^2\ln(s_0-p)\right] = Vt+\alpha$$

After introduction of the condition $p = 0$ when $t = 0$, this equation readily simplifies to

$$Vt = (1-K_m/K_p)(s_0-s) + \frac{(s_0-s)^2}{2K_p'} + K_m(1+s_0/K_p)\ln(s_0/s)$$

This equation differs from equation 8.8 only by the appearance of the term $\tfrac{1}{2}(s_0-s)^2/K_p'$, which disappears if $K_p' \to \infty$. The cases of pure non-competitive and uncompetitive inhibition can be generated by inserting the conditions $K_p \to \infty$ and $K_p = K_p'$, respectively, into this equation.

Because of the square term, the mixed inhibition case does not give a straight line if $t/\ln(s_0/s)$ is plotted against $(s_0-s)/\ln(s_0/s)$:

$$\frac{t}{\ln(s_0/s)} = \frac{1}{V}\left(1 - \frac{K_m}{K_p} + \frac{s_0-s}{2K_p'}\right)\frac{s_0-s}{\ln(s_0/s)} + \frac{K_m}{V}\left(1 + \frac{s_0}{K_p}\right) \quad (8.12)$$

because the slope is no longer a constant, but changes with s. Instead, the plots will be curved, as shown in *Figure 8.2*. However, the curvature is relatively slight unless K_p' is small compared with K_p, and may only become obvious in practice if the reaction is followed to 90% completion or more. This has two important consequences: (*i*) the extrapolation to obtain the initial velocities can be made accurately even if the inhibition is wrongly taken to be competitive when it is in fact mixed; (*ii*) significant curvature may pass unnoticed and give very inaccurate estimates of K_p. This means, on the one hand, that the occurrence of mixed inhibition provides no obstacle to the accurate estimation of V and K_m; but it also means, on the other hand, that K_p should not be estimated as described unless there is independent evidence that the inhibition is competitive. In addition, the apparent straightness of plots of $t/\ln(s_0/s)$ against $(s_0-s)/\ln(s_0/s)$ should not be taken as evidence of the nature of product inhibition. This, of course, can readily be established by the method described in the previous section for determining individual product-inhibition constants, i.e. by adding various amounts of product to the reaction mixture.

In general, the shapes of progress curves provide a rather insensitive test of whether particular equations are obeyed. In other words, it is often possible, not only in the specific situation that has been discussed, for an incorrect equation to describe an experimental progress curve closely. It is always advisable, therefore, to carry out several experiments at different initial substrate concentrations, just as would be done in initial rate studies. Nonetheless, it is wrong to conclude (as some workers have done) that integrated rate equations have no advantages over differentiated equations: a thorough analysis of a progress curve will *always* yield more information

148

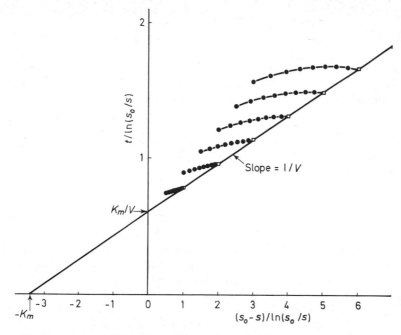

Figure 8.2 Effect of mixed inhibition by products on the method illustrated in Figure 8.1: Conditions are exactly as in Figure 8.1, with the same constants, but introducing an extra constant $K'_p = 8.27$ and applying equation 8.12 instead of equation 8.9. The extrapolated points (open squares) are unaffected by this change, because extrapolation to zero extent of reaction eliminates the effects of product

than an estimate of the initial rate from the same data. Even if the integrated equation is used only to estimate the initial rate, this estimate, if carried out properly, should always be more accurate than an estimate of initial slope, because it uses more information and is less affected by subjective bias. For a related method of estimating initial velocities, based on the direct linear plot, *see* Cornish-Bowden (1975).

8.6 More complex cases

Although we have considered only the relatively simple case of an irreversible single-substrate reaction, the principles discussed can be extended to most of the important cases in enzyme kinetics. Most of the relevant rate equations can be integrated much more readily than is commonly realized, requiring little more knowledge of calculus than the ability to look up standard integrals in a table. Alberty and Koerber (1957) examined the integrated form of the reversible Michaelis–Menten equation and applied it to fumarase; Schwert (1969) gave the integrated forms of the rate equations for a number of more complex mechanisms. Although the integration of these equations is not difficult, the solutions are generally rather complex, so that rather cumbersome procedures are required in order to analyse them. Moreover, integrated equations have been so little used in enzyme kinetics that it is probably unrealistic to pursue the subject further in detail.

In cases when it is not convenient to use the integrated rate equation (if, for example, the mechanism has not been established, or the enzyme is not stable during the course of the reaction), it is still possible to use progress curve data advantageously: instead of fitting the data to the true equation, they can be fitted instead to an equation of the form

$$p = \beta_0 + \beta_1 t + \beta_2 t^2 + \beta_3 t^3 + \cdots \tag{8.13}$$

which can be shown to define *any* single-valued continuous function. The coefficients $\beta_0, \beta_1, \beta_2, \beta_3 \ldots$, can be estimated by the method of least squares (Chapter 10), and then the initial velocity can be estimated from

$$\frac{dp}{dt} = \beta_1 + 2\beta_2 t + 3\beta_3 t^2 + \cdots \tag{8.14}$$

as $v = \beta_1$ when $t = 0$. In principle, the more terms are included in an equation of this type, the better the fit becomes. However, in practice, with enzymic progress curves almost all of the useful information is contained in the first three or four terms of equation 8.13 and the values of β_4 and higher terms are determined largely by random error. Consequently, it is never advisable to go beyond the term in t^3 and for many purposes the term in t^2 will define the curvature accurately enough (cf. Knowles, 1965).

The estimation of v_0 by fitting the progress curve to a power series in t has both advantages and disadvantages when compared with the use of integrated equations. On the one hand, it can be done regardless of whether the correct rate equation is known or not, and is not affected by complications such as progressive denaturation of the enzyme during the assay. On the other hand, it provides much less information about the curve. The coefficients $\beta_0, \beta_1, \beta_2 \ldots$ have no physical meaning, and cannot be transformed into physically meaningful parameters. In particular, they are *not* equal to the corresponding coefficients of the true equation expressed as an infinite series (this relationship holds only if both series contain an infinite number of terms). Although it might seem that substitution of a series of values of t into equations 8.13 and 8.14 would provide a series of (p, v) pairs that could be converted into (s, v) pairs to be analysed by using ordinary rate equations, this would not be a valid procedure. Pairs generated in this way would not be statistically independent, and could not therefore be subjected to statistical analysis.

8.7 Some pitfalls

The analysis of progress curves is fraught with traps for the unwary, and it is perhaps for this reason that so many enzymologists have taken the safe but unadventurous course of confining their attention to initial rates. In this section, some of the more serious errors are examined.

In the simple cases that have been considered in this chapter, the occurrence of competitive product inhibition has not affected the form of the integrated rate equation. Consequently, if competitive product inhibition is wrongly neglected, analysis of a single progress curve gives no indication of an error. This is also true in more complex cases because, whenever competitive product inhibition occurs, it manifests itself as a term in p in the denominator of the rate equation. A negative term in p must already be present, however, because the substrate concentration always occurs in the denominator as a term in $(s_0 - p)$. Thus, the product inhibition term does not affect the form of the integrated rate equation unless it equals or exceeds the term already present. It is profitable to consider this result in its converse form: *the observation that a progress curve is consistent with the absence of product inhibition provides no evidence that competitive product inhibition is absent.* For this reason, product inhibition should *always* be assumed to occur unless it has been shown not to occur. A simple method of checking this is to carry out experiments at several different values of s_0, as discussed in Section 8.3. It is worth emphasizing that product inhibition can be significant in a reaction that is, for all practical purposes, irreversible (cf. Section 2.7). Thus the knowledge of a very large equilibrium constant does not justify neglecting product inhibition.

A very easy error to make is to suppose that by estimating the slope at various positions along a progress curve one can generate a set of velocities, to be analysed as if they were initial rates estimated at various different values of s and p. For example, s might be determined at various points, and the velocity at the mid-point between each pair of points might be estimated by the difference between s values divided by the time difference. This is actually a generalization of what was once a very common erroneous method of estimating the parameters of straight-line plots: suppose that a set of equally spaced observations $y_1, y_2, y_3, \ldots, y_n$ are made of a quantity y that is a linear function of x. Then a series of slopes, $(y_2 - y_1)/\Delta x$, $(y_3 - y_2)/\Delta x, \ldots$, $(y_n - y_{n-1})/\Delta x$ might be estimated, and then the mean taken. Unfortunately, however, the mean is given by:

$$\text{Estimated slope} = \frac{1}{n-1}\left(\frac{y_2 - y_1}{\Delta x} + \frac{y_3 - y_2}{\Delta x} + \cdots + \frac{y_n - y_{n-1}}{\Delta x}\right)$$

$$= \frac{y_n - y_1}{(n-1)\Delta x}$$

so that only the first and last values make any contribution to the answer. The progress curves encountered in enzyme kinetics cannot be treated as simply as this, but essentially the same conclusion applies, and in general it is true that very little weight is given to intermediate observations in methods of this type (Cornish-Bowden, 1972).

A similar procedure, which is objectionable for slightly different reasons, is to obtain a best-fit polynomial equation of the form of equation 8.13 and then to insert different values of t into equation 8.14 in order to obtain a set of velocities. Although this method uses all of the observations, it is still invalid because the resulting errors in the velocities are very highly correlated and non-random. All of the usual statistical methods for the analysis of data

151

require that the observations be independent. When they are not, the estimates of the kinetic parameters may not be seriously in error, but any estimates of the precision of the parameters will be meaningless. The polynomial method is actually a 'smoothing' technique and destroys most of the information about the scatter of the data. Any other smoothing method, such as simply drawing a smooth curve through the observed line or points, is open to the same objection. Whenever smoothing is carried out, by any method, it is not valid to take more than one value from each curve for use in any subsequent statistical analysis. It is natural to take the initial slope as this one value, but this is not essential.

Common sense indicates that if a curve is analysed by means of, say, 50 representative points, the precision of the analysis will not be greatly improved by taking 500 points from the same curve. It is unreasonable to expect the additional 450 points to yield significantly more information than was contained in the first 50. Yet, if kinetic parameters are estimated with 50 and 500 points from the same curve, and statistical formulae are then blindly applied in order to calculate standard errors, these will be found to be much smaller with 500 points than with 50. The paradox is resolved if it is remembered that statistical calculations require *independent* observations: if points are taken that are too close together, they cannot be independent. Unfortunately, the question of how many points should be taken from one curve cannot be answered simply. If data collection and analysis can be carried out automatically, the time required for computer analysis can be used in order to decide the number of points, because the inclusion of more points than necessary does not affect the accuracy of the parameter estimates. If data collection and analysis are carried out manually, however, it is obviously sensible to avoid unnecessary labour. In such cases, about 10 points are likely to be adequate to define most curves, and certainly not more than 20 are likely to be required. Even if only a few points are used, estimates of standard errors of parameters should be treated with even more scepticism than usual, because one has no idea how many of the observations can be considered to be independent. In any event, the variations between curves will normally be much greater than the variations between points in a single curve. It is sensible, if the method described in Section 8.3 is used in order to obtain values of the initial velocities, to regard each extrapolated initial velocity as a single observation and each slope as a single observation in the estimation of K_p.

In spite of the various problems that have been discussed in this section, there is no doubt that the analysis of progress curves forms a very valuable part of enzyme kinetics, provided that it is carried out cautiously and sensibly.

9
Fast Reactions

9.1 Limitations of steady-state measurements

It is convenient to refer to the period in a reaction before the steady state is reached as the *transient phase*. This term is commonly used in physics and mathematics to describe terms of the form $A \exp(-t/\tau)$ that often occur in the solutions of differential equations. Such terms have finite and even very large values when t is small, but decay to zero as t is increased above τ, a constant called the '*relaxation time*.' As will be seen, they always occur in kinetic equations if the steady-state assumption is not made during the derivation.

It is fairly obvious that experimental methods for investigating very fast reactions, with half-times of much less than 1 s, must be different from those used for slower reactions, because in most of the usual methods the time taken in mixing the reactants is of the order of seconds or greater. Although mechanical mixing devices, such as the 'stopped-flow' apparatus, permit more or less conventional methods to be used in the study of reactions with half-times as short as 10^{-3} s, faster reactions require different methods. It is rather less obvious that the kinetic equations required for fast reactions are also different, because in most enzyme-catalysed reactions the steady state is attained very rapidly and can be considered to exist throughout the period of investigation, provided that this period does not include the first second after mixing. Consequently, most of the equations that have been discussed in this book have been derived with the use of the steady-state assumption. However, fast reactions are concerned, almost by definition, with the transient phase before the attainment of a steady state and cannot be described by steady-state rate equations. This chapter is concerned with the derivation of rate equations for the transient phase, but first it is useful to examine the reasons for making transient-phase measurements.

Although steady-state measurements have proved extremely useful in elucidating the mechanisms of enzyme-catalysed reactions, they suffer from the major disadvantage that, at best, the steady-state velocity of a multi-step

153

reaction is the velocity of the slowest step, and steady-state measurements do not normally provide information about any of the faster steps. Yet if the mechanism of an enzyme-catalysed reaction is to be understood, it is necessary to have information about steps other than the slowest. As discussed in Chapter 5, the experimenter has considerable freedom to alter the relative rates of the various steps in a reaction, by varying the concentrations of the substrates. Consequently, it is often possible to examine more than one step of a reaction in spite of this limitation. However, isomerizations of intermediates along the reaction pathway cannot be separated in this way. To take a simple example:

$$E + A \underset{k_{-1}}{\overset{k_{+1}}{\rightleftarrows}} EA \underset{k_{-2}}{\overset{k_{+2}}{\rightleftarrows}} EP \underset{k_{-3}}{\overset{k_{+3}}{\rightleftarrows}} E + P$$

The steady-state equation for this mechanism is equation 2.18, i.e.

$$v = \frac{(V^f a / K_m^A) - (V^r p / K_m^P)}{1 + (a / K_m^A) + (p / K_m^P)}$$

which contains only four parameters, and it is impossible to obtain from them the values of all of the six independent rate constants. This equation also applies to much more complex mechanisms where the interconversion of EA and EP involves several intermediates. Steady-state measurements not only fail to provide any information about the individual steps, but also give no indication of how many steps there are. In general, as mentioned in Section 3.3, in any part of a reaction pathway that consists of a series of isomerizations of intermediates, all of the intermediates must be treated as a single species in steady-state kinetics. This is a severe limitation and provides the main justification for transient-state kinetics, which are subject to no such limitation.

The advantages of transient-state methods may seem to make steady-state kinetics obsolete, but it is likely that steady-state investigations will continue to predominate for many years, for reasons that will now be considered. Firstly, the theory of the steady state is simpler, and steady-state measurements require less expensive equipment. In addition, steady-state measurements need very small amounts of enzyme and, although this may be a disadvantage if it means that conditions in the assay mixture are very different from those in the cell, it will continue to be a deciding factor until methods of purifying enzymes have advanced far beyond their present level of convenience and efficiency. It is significant in this connection that most transient-state studies have been carried out on enzymes of little metabolic interest, such as chymotrypsin, papain, ficin and lysozyme, because they are readily available in large amounts, and until such studies have been applied routinely to enzymes such as phosphofructokinase and glutamine synthetase, for example, it will be premature to regard the steady state as obsolete.

One should also be aware that analysis of transient-state data suffers severely from a numerical difficulty known as *ill-conditioning*. This means that, *even in the absence of experimental error*, it is possible to fit experimental results with a wide range of constants and indeed of equations. This is illustrated in *Figure 9.1*, which shows a set of points and a line calculated

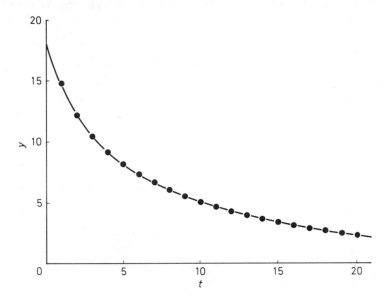

Figure 9.1 Ill-conditioned character of exponential functions: The points were calculated from $y = 5.1\ exp(-t/1.3) + 4.7\ exp(-t/4.4) + 9.3\ exp(-t/14.2)$, *the line from* $y = 7.32\ exp(-t/2.162) + 10.914\ exp(-t/12.86)$

from two different equations, both of the type commonly encountered in transient-state kinetics. The practical implication is that it is often impossible to extract all of the extra information that is theoretically present in transient-state measurements unless the various processes are very well separated on the time scale.

In conclusion, all enzymologists should be aware of the potential usefulness of transient-state methods, but they should not expect to abandon the steady state entirely.

9.2 Transient phase of the Michaelis–Menten mechanism

In Section 2.4, the rate equation for the Michaelis–Menten mechanism was derived without making the steady-state assumption, in order to demonstrate that under appropriate conditions a steady state would in fact be reached very rapidly. The transient phase of this mechanism will now be examined in more detail. It was found (equation 2.9) that the rate could be represented by:

$$v = \frac{\mathrm{d}p}{\mathrm{d}t} = \frac{k_{+1}k_{+2}e_0s\{1 - \exp[-(k_{+1}s + k_{-1} + k_{+2})t]\}}{k_{+1}s + k_{-1} + k_{+2}} \tag{9.1}$$

155

It is plain from this equation that the rate is initially zero but increases rapidly to the steady-state value as the exponential term decays. The actual dependence of p on t can be found by integrating and introducing the condition $p = 0$ when $t = 0$:

$$p = \frac{k_{+1}k_{+2}e_0 s t}{k_{+1}s + k_{-1} + k_{+2}} - \frac{k_{+1}k_{+2}e_0 s\{1 - \exp[-(k_{+1}s + k_{-1} + k_{+2})t]\}}{(k_{+1}s + k_{-1} + k_{+2})^2} \quad (9.2)$$

The plot of p against t, illustrated in *Figure 9.2*, initially curves upwards, but

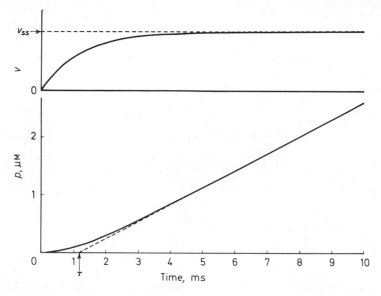

Figure 9.2 Approach to the steady state: In the upper part of the figure, the velocity v *is plotted against time and the broken line shows the steady-state velocity* v_{ss}. *In the lower part of the figure, the concentration of product,* p, *is plotted against time with the same time scale. The values were calculated from equations 9.1 and 9.2, with* $k_{+1} = 50\,000\ \text{M}^{-1}\ \text{s}^{-1}$, $k_{-1} = 500\ \text{s}^{-1}$, $k_{+2} = 100\ \text{s}^{-1}$, $s = 0.005$ M *and* $e_0 = 0.000\,01$ M

as the exponential term decays it becomes a straight line given by the equation

$$p = \frac{k_{+1}k_{+2}e_0 s}{k_{+1}s + k_{-1} + k_{+2}} \left(t - \frac{1}{k_{+1}s + k_{-1} + k_{+2}} \right) \quad (9.3)$$

Extrapolation of this line back to the t axis gives an intercept, τ, of $1/(k_{+1}s + k_{-1} + k_{+2})$. As the value of τ is dependent on s, a plot of $1/\tau$ against s gives a straight line of slope k_{+1} and intercept $(k_{-1} + k_{+2})$ on the $1/\tau$ axis, as was demonstrated by Gutfreund (1955) for the ficin-catalysed hydrolysis of benzoyl-L-arginine ethyl ester.

9.3 'Burst' kinetics

In studies of the chymotrypsin-catalysed hydrolysis of nitrophenylethyl carbonate, Hartley and Kilby (1954) observed that, although the release of nitrophenol was almost linear, extrapolation of the line back to the product axis gave a positive intercept. Because the substrate was very poor, it was necessary to use high enzyme concentrations and the intercept, which is known as a *'burst'* of product, was proportional to the enzyme concentration. This suggested a mechanism in which the products were released in two steps, the nitrophenol being released first:

$$E + S \underset{k_{-1}}{\overset{k_{+1}}{\rightleftharpoons}} ES \xrightarrow{k_{+2}} EQ \xrightarrow{k_{+3}} E + Q \qquad (9.4)$$
$$\searrow P$$

If the final step is rate limiting, i.e. if k_{+3} is small compared with $k_{+1}s$, k_{-1} and k_{+2}, then the enzyme will exist almost entirely as EQ in the steady state. However, it is not necessary for EQ to be formed before P can be released, and so in the transient phase P can be released at a rate much greater than the steady-state rate. It might be expected that the amount of P released in the burst would be equal, and not merely proportional, to the amount of enzyme. This is accurately true only if k_{+3} is very much smaller than the other rate constants; otherwise, the burst is smaller than the stoichiometric amount, as will now be shown, following a derivation based on that of Gutfreund (1955).

If s is large enough to be treated as a constant, and if $k_{+1}s$ is large compared with $(k_{-1}+k_{+2}+k_{+3})$, then very shortly after mixing the system effectively simplifies to

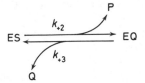

because the reaction $E + S \rightarrow ES$ can be regarded as instantaneous and irreversible, and the concentration of free enzyme becomes negligible. This is then a simple reversible first-order reaction (cf. Section 1.4), with the solution

$$[ES] = \frac{e_0\{k_{+3}+k_{+2}\exp[-(k_{+2}+k_{+3})t]\}}{k_{+2}+k_{+3}}$$

$$[EQ] = \frac{k_{+2}e_0\{1-\exp[-(k_{+2}+k_{+3})t]\}}{k_{+2}+k_{+3}}$$

From these equations, expressions for dp/dt and dq/dt are readily obtained:

$$\frac{dp}{dt} = k_{+2}[ES] = \frac{k_{+2}e_0\{k_{+3}+k_{+2}\exp[-(k_{+2}+k_{+3})t]\}}{k_{+2}+k_{+3}}$$

$$\frac{dq}{dt} = k_{+3}[EQ] = \frac{k_{+2}k_{+3}e_0\{1-\exp[-(k_{+2}+k_{+3})t]\}}{k_{+2}+k_{+3}}$$

157

In the steady state, i.e. when t is large, the exponential term becomes negligible and these two equations simplify to

$$\frac{dp}{dt} = \frac{dq}{dt} = \frac{k_{+2}k_{+3}e_0}{k_{+2}+k_{+3}}$$

However, in the transient phase dp/dt is initially much larger than dq/dt, so that while P displays a 'burst', Q displays a 'lag' if the linear parts of the curves are extrapolated back. The magnitudes of these intercepts can be calculated by integrating and introducing the conditions $p = 0$ and $q = 0$ when $t = 0$:

$$p = \frac{k_{+2}k_{+3}e_0 t}{k_{+2}+k_{+3}} + \frac{k_{+2}^2 e_0\{1-\exp[-(k_{+2}+k_{+3})t]\}}{(k_{+2}+k_{+3})^2} \tag{9.5}$$

$$q = \frac{k_{+2}k_{+3}e_0 t}{k_{+2}+k_{+3}} - \frac{k_{+2}k_{+3}e_0\{1-\exp[-(k_{+2}+k_{+3})t]\}}{(k_{+2}+k_{+3})^2} \tag{9.6}$$

The linear part of the plot of p against t is obtained by omitting the exponential term from equation 9.5, and extrapolation to zero time gives π, the magnitude of the burst:

$$\pi = \frac{k_{+2}^2 e_0}{(k_{+2}+k_{+3})^2} = \frac{e_0}{\left(1 + \dfrac{k_{+3}}{k_{+2}}\right)^2} \tag{9.7}$$

Thus the burst of P is *not* equal to the enzyme concentration, but approximates to it if $k_{+2} \gg k_{+3}$. This equation implies that the burst can never exceed the enzyme concentration, but extrapolation of the 'linear' portion of a progress curve can sometimes yield an overestimate of the true burst size if the velocity is not truly constant in the steady state but decays at a significant rate. One can avoid this type of error by ensuring that the progress curve is truly straight during the steady-state phase.

The lag, τ, in the production of Q can be found in a similar way, and is given by

$$\tau = 1/(k_{+2}+k_{+3})$$

τ is likely to be detectable only if k_{+2} is of similar magnitude to k_{+3}. The magnitude of this type of lag is independent of both enzyme and substrate concentrations, and so it can readily be distinguished from the type of lag that was discussed in the previous section.

A pair of progress curves for this mechanism are given in *Figure 9.3*, calculated with $k_{+2} = 10k_{+3}$, in order to illustrate these results.

A rather more rigorous analysis of the mechanism shown in equation 9.4 is possible, involving no assumptions about the relative magnitudes of the rate constants (Ouellet and Stewart, 1959). In the full solutions, $k_{+1}s_0$ and k_{-1} appear in the equations together with an additional exponential term, which gives rise to a very brief lag before the burst occurs. This analysis is of great theoretical interest, but it has not proved to be of wide practical application and will not be further discussed here.

The results discussed in this section have led to an important method for

158

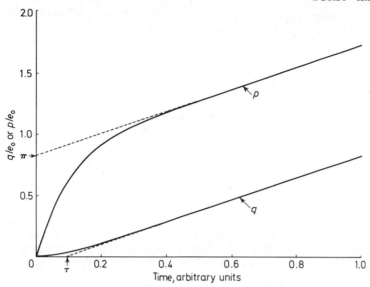

Figure 9.3 'Burst' kinetics: The concentrations of P and Q are shown as functions of time for an enzyme following the mechanism shown in equation 9.4, with $k_{+2} = 10$, $k_{+3} = 1$ (arbitrary units), under saturating conditions, i.e. $k_{+1}s \gg (k_{-1} + k_{+2} + k_{+3})$

titrating enzymes. It is generally difficult to obtain an accurate measure of the molarity of an enzyme: rate assays provide concentrations in activity units per millilitre, which are adequate for comparative purposes but are not true concentrations; most other assays are really protein assays and are therefore very unspecific. However, equation 9.7 shows that, if a substrate can be found for which k_{+3} is either very small or zero, then the burst π is equal to the enzyme concentration. The end-point is also very well defined if k_{+3} is very small. The substrates of chymotrypsin that were examined originally, p-nitrophenylethyl carbonate and p-nitrophenyl acetate, showed values of k_{+3} that were inconveniently large, but subsequently Schonbaum, Zerner and Bender (1961) found that under suitable conditions *trans*-cinnamoylimidazole gave excellent results. This compound reacts rapidly with chymotrypsin to give imidazole and *trans*-cinnamoylchymotrypsin, but no further reaction occurs readily. Measurement of the amount of imidazole released by a solution of chymotrypsin provides a measure of the amount of enzyme. Suitable titrants for a number of other enzymes have also been found (*see*, e.g. Bender *et al.*, 1966).

Active site titration by means of burst measurements differs from rate assays in being relatively insensitive to changes in the rate constants: a rate assay demands very well defined conditions of pH, temperature, buffer composition, etc., if it is to be reproducible, but the magnitude of a burst is unaffected by relatively large changes in k_{+2}, such as might result from chemical modification of the enzyme, unless these reduce k_{+2} to a very small value. Thus chemical modification alters the molarity of an enzyme, as measured

159

by this technique, either to zero or not at all. For this reason, enzyme titration has also been called an '*all-or-none*' assay (Koshland, Strumeyer and Ray, 1962).

9.4 Reversible sequences of reactions

There are two types of mechanism for which exact solutions to the rate equation exist. One of these is the reversible bimolecular reaction, exemplified by

$$E + S \rightleftarrows ES$$

This mechanism was considered in Section 1.4. It is of rather limited application in biochemistry, because few reactions are as simple as this. Even for reactions that do correspond to this mechanism, it is more usual to apply the approximate solutions discussed in the following sections, because these are very widely applicable. However, the full solution to this mechanism has been used occasionally, as, for example, by Ellis and Dunford (1968) for the binding of cyanide to peroxidase.

The other type of mechanism for which an exact solution exists is a sequence of n reversible unimolecular steps:

$$X_0 \rightleftarrows X_1 \rightleftarrows X_2 \rightleftarrows \cdots \rightleftarrows X_n \tag{9.8}$$

The limitation to unimolecular steps is not as restrictive as it might seem because, although all enzyme-catalysed reactions involve at least one bimolecular step, it is usually possible to achieve pseudo-first-order conditions. The rate equations that describe the following generalized enzyme mechanism:

$$E + S \rightleftarrows ES_1 \rightleftarrows ES_2 \rightleftarrows \cdots \rightleftarrows ES_n \rightleftarrows E + P$$

can be solved with good accuracy if either of the following conditions holds:
  (1) If s is effectively constant, and p is effectively zero, then the first step, $E + S \rightarrow ES_1$, is pseudo-first order and the first step in the reverse direction, $E + P \rightarrow ES_n$, can be ignored. In this case, equation 9.8 applies.
  (2) If a steady state exists, the steady-state rate equations can be integrated as described in Chapter 8.

One or other of these conditions applies at all stages during most enzyme-catalysed reactions. The solution for the transient phase (condition 1) invariably contains terms of the form $\exp(-t/\tau)$. Provided that the largest of these terms becomes insignificant before the substrate concentration changes significantly, the transient and steady-state phases can be considered separately, ignoring changes in s during the transient phase and assuming that changes in the concentrations of intermediates are very slow during the steady-state phase.

A third condition also permits accurate solution of the rate equations: if a reaction is close to equilibrium, all of the non-linear terms in the rate equations involve products of small numbers, and can be ignored with good accuracy. The solutions resemble those for the transient phase and provide similar information, but are often much more convenient experimentally. This situation is discussed in the latter part of this chapter.

Although equation 9.8 can be treated exactly for any number of steps, it is easier to follow the derivation for a specific example:

$$X_0 \underset{k_{-1}}{\overset{k_{+1}}{\rightleftharpoons}} X_1 \underset{k_{-2}}{\overset{k_{+2}}{\rightleftharpoons}} X_2 \tag{9.9}$$

The system is defined by a conservation equation:

$$x_0 + x_1 + x_2 = x_{tot} \tag{9.10}$$

and three rate equations:

$$dx_0/dt = -k_{+1}x_0 + k_{-1}x_1 \tag{9.11}$$

$$dx_1/dt = k_{+1}x_0 - (k_{-1} + k_{+2})x_1 + k_{-2}x_2 \tag{9.12}$$

$$dx_2/dt = k_{+2}x_1 - k_{-2}x_2 \tag{9.13}$$

Any one of the three rate equations is redundant, as their sum is the same as the first derivative of equation 9.10, i.e. $dx_0/dt + dx_1/dt + dx_2/dt = 0$, so there are in fact three independent linear differential equations in three unknowns. Before they can be solved, they must be converted into a single differential equation in one unknown. x_2 can be eliminated between equations 9.10 and 9.12:

$$dx_1/dt = k_{+1}x_0 - (k_{-1} + k_{+2})x_1 + k_{-2}(x_{tot} - x_0 - x_1)$$
$$= k_{-2}x_{tot} + (k_{+1} - k_{+2})x_0 - (k_{-1} + k_{+2} + k_{-2})x_1 \tag{9.14}$$

Differentiation of equation 9.11 gives

$$d^2x_0/dt^2 = -k_{+1}\,dx_0/dt + k_{-1}\,dx_1/dt \tag{9.15}$$

dx_1/dt is eliminated by substituting equation 9.14 in equation 9.15:

$$d^2x_0/dt^2 = -k_{+1}\,dx_0/dt + k_{-1}k_{-2}x_{tot}$$
$$+ (k_{+1} - k_{-2})k_{-1}x_0 - k_{-1}(k_{-1} + k_{+2} + k_{-2})x_1 \tag{9.16}$$

Finally, x_1 is eliminated between equations 9.11 and 9.16:

$$d^2x_0/dt^2 + (k_{+1} + k_{-1} + k_{+2} + k_{-2})dx_0/dt$$
$$+ (k_{-1}k_{-2} + k_{+1}k_{+2} + k_{+1}k_{-2})x_0 = k_{-1}k_{-2}x_{tot}$$

This is of the standard form $d^2x_0/dt^2 + P\,dx_0/dt + Qx_0 = R$, and has the solution

$$x_0 = A_1 \exp(-t/\tau_1) + A_2 \exp(-t/\tau_2)$$

where A_1 and A_2 are constants of integration defined by the initial state of the system, and τ_1 and τ_2 are given by:

$$1/\tau_1 = \tfrac{1}{2}(P + \sqrt{P^2 - 4Q})$$

$$1/\tau_2 = \tfrac{1}{2}(P - \sqrt{P^2 - 4Q})$$

161

If $k_{-1}k_{+2}$ is small compared with $(k_{-1}k_{-2}+k_{+1}k_{+2}+k_{+1}k_{-2})$, these solutions simplify to the following convenient values:

$$1/\tau_1 = k_{+1}+k_{-1} \qquad (9.17)$$

$$1/\tau_2 = k_{+2}+k_{-2} \qquad (9.18)$$

For the general case of a mechanism of n unimolecular steps, all except one of the unknown concentrations can be eliminated in a similar way to that described for equation 9.9. The conservation equation is used in order to eliminate the first unknown, but each subsequent elimination involves differentiation with respect to t, so that the resulting differential equation in one unknown is an nth order linear differential equation, with a solution containing n exponential terms. In favourable cases, some of the relaxation times can be associated with specific steps, as in equations 9.17 and 9.18, but this simplification does not apply generally.

9.5 Jump kinetics

One of the most useful techniques for studying rapid reactions is to observe the relaxation of a system back to equilibrium after a small perturbation. By far the most important in enzyme kinetics is the *temperature-jump* method, in which the perturbation of equilibrium is the result of a sudden increase in temperature. With relatively simple apparatus, a temperature increase of $10\,°C$ in 10^{-6} s is possible. As most kinetic parameters vary with temperature, a temperature jump is normally followed by a rapid relaxation to a new equilibrium. Other perturbations are possible, such as a pressure jump, but these are less useful because they produce much smaller changes in the kinetic parameters for a comparable input of energy. However, the same analysis applies to any type of perturbation.

A simple binding reaction will be used as an example:

$$E \;+\; S \;\underset{k_{-1}}{\overset{k_{+1}}{\rightleftarrows}}\; ES$$

$$e_\infty+\Delta e \qquad s_\infty+\Delta s \qquad x_\infty+\Delta x$$

If e_∞, s_∞ and x_∞ are the final equilibrium concentrations of E, S and ES, respectively, and k_{+1} and k_{-1} are the rate constants after the perturbation, then the instantaneous concentrations can be represented as $(e_\infty+\Delta e)$, $(s_\infty+\Delta s)$ and $(x_\infty+\Delta x)$, respectively. Hence the rate is given by

$$dx/dt = k_{+1}(e_\infty+\Delta e)(s_\infty+\Delta s)-k_{-1}(x_\infty+\Delta x)$$

But $d\Delta x/dt = dx/dt$ and, by the stoichiometry of the reaction, $\Delta e = \Delta s = -\Delta x$, and so

$$d\Delta x/dt = k_{+1}(e_\infty-\Delta x)(s_\infty-\Delta x)-k_{-1}(x_\infty+\Delta x) \qquad (9.19)$$

As there is no net rate at equilibrium, $k_{+1}e_\infty s_\infty = k_{-1}x_\infty$, and $(\Delta x)^2$ can be neglected if Δx is small, so equation 9.19 simplifies to

162

$$d\Delta x/dt = -[k_{+1}(e_\infty + s_\infty) + k_{-1}]\Delta x$$

This is a simple linear differential equation and can be solved to give

$$\Delta x = \Delta x_0 \exp\{-[k_{+1}(e_\infty + s_\infty) + k_{-1}]t\}$$

where Δx_0 is the initial perturbation. Thus, provided that the initial perturbation is small, the relaxation of a single-step reaction is described by a single exponential term with a relaxation time, τ, given by

$$1/\tau = k_{+1}(e_\infty + s_\infty) + k_{-1}$$

As e_∞, s_∞ and k_{-1}/k_{+1} can normally be measured independently, measurement of τ permits individual values to be assigned to k_{+1} and k_{-1}.

Mechanisms that involve several steps can be treated in exactly the same way, although the derivation is more complicated. In general, the solution for an n-step mechanism is characterized by n relaxation times, unless the steps are related by thermodynamic requirements, as in

$$\begin{array}{ccc} E & + \; S \rightleftarrows & ES \\ \updownarrow & & \updownarrow \\ E' & + \; S \rightleftarrows & E'S \end{array}$$

Here, although there are four equilibria, only three are independent because the fourth equilibrium constant is determined completely by the other three. In such cases, the number of distinct relaxation times is equal to the number of independent equilibria. Mechanisms of this type occur frequently in the treatment of proton binding.

An important type of multi-step mechanism is one in which the initial bimolecular reaction is followed by several unimolecular isomerization steps:

$$E + S \rightleftarrows ES \rightleftarrows E'S \rightleftarrows E''S \ldots$$

The expressions for the relaxation times are the same as those for the sequence of unimolecular steps discussed in Section 9.4, except that the pseudo-first-order rate constant for the bimolecular step, $E + S \rightarrow ES$, is $k_{+1}(e_\infty + s_\infty)$ instead of simply k_{+1}.

Although the jump method has been discussed in terms of relaxation to equilibrium, the same analysis can be applied equally well to relaxation to a steady state:

$$E \quad + \quad S \; \underset{k_{-1}}{\overset{k_{+1}}{\rightleftharpoons}} \; ES \; \overset{k_{+2}}{\longrightarrow} \; E \quad + \quad P$$
$$e_{ss} + \Delta e \quad s_0 \qquad\qquad x_{ss} + \Delta x$$

The relaxation rate is given by

$$d\Delta x/dt = k_{+1}(e_{ss} + \Delta e)s_0 - (k_{-1} + k_{+2})(x_{ss} + \Delta x)$$

After introducing the steady-state condition $k_{+1}e_{ss}s_0 = (k_{-1} + k_{+2})x_{ss}$, and the stoichiometry requirement $\Delta e = -\Delta x$, this simplifies to

$$d\Delta x/dt = -(k_{+1}s_0 + k_{-1} + k_{+2})\Delta x$$

163

and has the solution

$$\Delta x = \Delta x_0 \exp(-t/\tau)$$

where the relaxation time τ, given by

$$1/\tau = k_{+1}s_0 + k_{-1} + k_{+2}$$

is the same as the relaxation time for the transient phase after rapid mixing of reactants, discussed in Section 9.2 for the same mechanism.

Experimentally, the important difference between the rapid-mixing and temperature-jump methods is in the range of relaxation times that can be measured. Even the most efficient mixing devices require at least 10^{-3} s for complete mixing, and cannot be used to measure shorter relaxation times. In the temperature-jump method, the reactants are already mixed before the perturbation, and the limiting factor is the time required for heating (about 10^{-6} s). Thus much shorter relaxation times can be measured. On the other hand, convection effects prevent the temperature-jump method from being used for relaxation times greater than about 10^{-2} s, so it cannot replace the rapid-mixing methods completely.

The temperature-jump and other relaxation methods were proposed by Eigen (1954), and have been applied by him and co-workers to a wide range of systems in chemistry and biochemistry. The first application to enzyme kinetics occurred when Hammes and Fasella (1962, 1963) used the temperature-jump method to study the interaction of various compounds with glutamate–aspartate transaminase; and in recent years Hammes and co-workers have used relaxation techniques in the study of numerous enzymes (see Hammes, 1968; Hammes and Schimmel, 1970).

9.6 Sinusoidal perturbations

In discussing jump kinetics, the perturbation has been implicitly regarded as instantaneous, but a more realistic approximation is to consider each rate constant, k, to suffer an exponential change:

$$k = k_\infty[1 - A \exp(-t/\tau)]$$

where τ is of the order of 10^{-6} s and A is determined by the magnitude of the perturbation. This equation simplifies to the 'instantaneous' approximation, $k = k_\infty$, if the period of study is confined to values of $t > 10^{-6}$ s. Very much faster reactions can be studied with sinusoidal perturbations. These are commonly produced by passing ultrasonic waves, of frequency as high as 10^{14} s^{-1}, through the reaction mixture. Ultrasonic waves are accompanied by local fluctuations in temperature and pressure as they are transmitted through a medium. In water, the temperature fluctuations are very small, but they have more effect than the pressure fluctuations on the kinetic parameters, because pressure changes have very little effect on reactions in solution. However, as both fluctuations are exactly in phase, it is of no consequence which of them produces the greater effect. Provided that the perturbations are small, they produce proportionate changes in the rate constants and so, for a simple binding reaction:

$$E + S \underset{k_{-1}}{\overset{k_{+1}}{\rightleftharpoons}} ES$$

$$e \qquad s \qquad x$$

each rate constant varies sinusoidally in response to the sinusoidal ultrasonic wave:

$$k_{+1} = \overline{k}_{+1}(1 + A \sin \omega t) \qquad (9.20)$$

$$k_{-1} = \overline{k}_{-1}(1 + B \sin \omega t) \qquad (9.21)$$

where $\overline{k}_{+1}$ and $\overline{k}_{-1}$ are the mean values of k_{+1} and k_{-1}, ω is the frequency of the ultrasonic wave and A and B are the amplitudes. The rate equation is

$$dx/dt = k_{+1}es - k_{-1}x \qquad (9.22)$$

If $\bar{e}$, $\bar{s}$ and $\bar{x}$ are the unperturbed values of e, s and x, then stoichiometry requires that

$$e - \bar{e} = s - \bar{s} = -(x - \bar{x}) = -\Delta x \qquad (9.23)$$

and substitution of equations 9.20, 9.21 and 9.23 into equation 9.22 gives

$$d\Delta x/dt = \overline{k}_{+1}(1 + A \sin \omega t)(\bar{e} - \Delta x)(\bar{s} - \Delta x) - k_{-1}(1 + B \sin \omega t)(\bar{x} + \Delta x) \quad (9.24)$$

Now, $\overline{k}_{+1}\bar{e}\bar{s} = \overline{k}_{-1}\bar{x}$ if the unperturbed system is at equilibrium (or a steady state), and Δx, A and B are all small, so $(\Delta x)^2$, $A\Delta x$ and $B\Delta x$ can be neglected. Therefore, equation 9.24 simplifies to:

$$d\Delta x/dt + [\overline{k}_{+1}(\bar{e} + \bar{s}) + \overline{k}_{-1}]\Delta x = \overline{k}_{+1}\bar{e}\bar{s}(A - B)\sin \omega t$$

which is a standard differential equation, with the solution

$$\Delta x = \frac{\overline{k}_{+1}\bar{e}\bar{s}(A - B)\left(\dfrac{1}{\tau} \sin \omega t - \omega \cos \omega t\right)}{\dfrac{1}{\tau^2} + \omega^2} + R \exp(-t/\tau) \qquad (9.25)$$

where τ is given by

$$\frac{1}{\tau} = \overline{k}_{+1}(\bar{e} + \bar{s}) + \overline{k}_{-1}$$

and R is a constant of integration. The exponential term becomes negligible when $t \gg \tau$, and equation 9.25 simplifies to

$$\Delta x = \frac{\overline{k}_{+1}\bar{e}\bar{s}(A - B)(\tau \sin \omega t - \omega\tau^2 \cos \omega t)}{1 + \omega^2\tau^2}$$

Comparison with an equation showing the variation of the (unattained) equilibrium displacement, Δx_{eqm}:

$$\Delta x_{eqm} = \overline{k}_{+1}\bar{e}\bar{s}(A - B)\tau \sin \omega t$$

165

shows that the actual displacement oscillates with the same frequency as the equilibrium displacement, but lags behind it by $\tan^{-1}(\omega\tau)$ rad and has a reduced amplitude, a, given by

$$a = \frac{a_{eqm}}{(1+\omega^2\tau^2)^{\frac{1}{2}}}$$

where a_{eqm}, given by

$$a_{eqm} = \overline{k}_{+1}\overline{es}(A-B)\tau$$

is the amplitude of the equilibrium displacement. This is illustrated in *Figure 9.4* for various values of $\omega\tau$. When $\omega\tau$ is small, the reaction remains

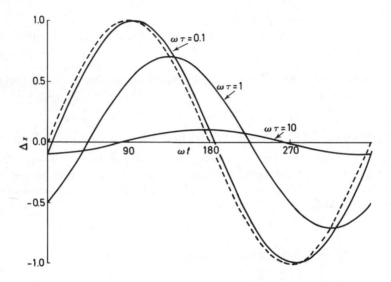

Figure 9.4 Response of a binding reaction $E + S \rightleftarrows ES$ to sinusoidal perturbation of the rate constants: The broken line shows the variation of the unattained equilibrium response, and the solid lines show the actual responses for three values of $\omega\tau$, as indicated

close to equilibrium at all times, but as the frequency and $\omega\tau$ increase it lags further behind and the amplitude decreases, and at very high frequencies there is hardly any response.

As chemical reactions have no natural tendency to oscillate about their equilibria, power must be absorbed from the input ultrasonic wave to force the chemical oscillations. This power is proportional to the squares of both the amplitude and the frequency of oscillation. Hence the power absorbed per wavelength is proportional to $a_{eqm}^2\omega/(1+\omega^2\tau^2)$. This function has a maximum at a point where $\omega = 1/\tau$, and so measurement of the power absorbed at different frequencies provides a simple method of measuring τ.

The ultrasonic absorption method has been used to study numerous simple reactions, such as the solvation of glycine and related compounds (Hammes

166

and Pace, 1968). Investigations of enzymes have been hindered by experimental difficulties, such as the need for very large amounts of material in order to obtain significant absorption, and by the daunting theoretical complexity of the expected relaxation spectrum. However, there can be no doubt that ultrasonic absorption will eventually become a most valuable technique for studying very fast steps in enzyme-catalysed reactions. It has been used in the study of conformational changes in poly-L-glutamic acid (Burke, Hammes and Lewis, 1965), with results of obvious relevance to enzyme kinetics. This study, and studies of simple systems, are important in understanding enzyme catalysis because they define 'reasonable' values for the rates of many elementary processes, such as solvation, protonation, deprotonation and macromolecular conformational changes. Thus, although the observation that glycine solvation occurs with a relaxation time of about 2.5×10^{-9} s does not prove that solvation occurs equally fast in all reactions, it does provide an estimate of the rate to be expected in other systems. Similarly, the determination of the relaxation time for the helix–coil transition in poly-L-glutamic acid, in the range 5×10^{-8} to 10^{-5} s, provides an estimate of the rates that are possible for conformational changes in enzyme-catalysed reactions.

10
Estimation of Rate Constants

10.1 Value and limitations of a statistical approach

For the last 40 years, most enzyme kinetic experiments have been analysed by means of linear plots of the type described in Section 2.5. The double-reciprocal plot, in which $1/v$ is plotted against $1/s$, has been by far the most widely used plot but it is also, unfortunately, the most objectionable from the statistical point of view and provides very poor estimates of K_m and V. Despite advances in analytical methods during recent years, the popularity of the double-reciprocal plot continues unabated, partly because some leading enzymologists have recommended it without fully understanding its faults, and partly because it is (very slightly) easier to use than most other methods.

In definitive work, it is preferable to avoid all plots and to use statistical analysis instead, but many experimentalists have found the statistical arguments rather arid, believing that the advantages to be gained from valid statistical analysis are too slight to off-set the loss in convenience. It is therefore appropriate to introduce this chapter with a discussion of the value and limitations of statistical analysis in general.

If all experiments were executed with perfect precision, there would be no objection to graphical methods of analysis. The only limitation would be the ability of the experimenter to plot the data precisely and to read off the results precisely. However, real experiments are always subject to experimental error, and the main aim of statistical analysis is to minimize its deleterious effects. Graphical methods do not, in general, achieve this end, because it is impossible for an experimenter to give correct weight to all observations when drawing a line by eye, particularly if the correct weights vary from observation to observation. In practice, any line drawn by eye through a series of points is subjectively biassed. This is objectionable because a biassed line provides less accurate estimates than an unbiassed line, and because any bias in technique reduces the credibility of the conclusions that an experimenter may wish to draw.

A second objection to graphical methods is that usually they do not provide any information about the precision of the estimated kinetic constants (although the direct linear plot is an exception: *see* Section 10.9). Statistical calculations are much superior, in that they provide estimates of the errors that result from random variations in the observations.

It is instructive to consider the various types of error that can arise in an experiment. Random variations are of two types, experimental error and sampling variation. True experimental error results from inaccuracies in pipetting, instrumental noise, imprecise reading of values from chart recorders and similar sources. Many enzymologists regard experimental error as the only important source of random variations in enzyme kinetics, but some, e.g. Reich (1970), believe that sampling variation is also important. In some fields it is overwhelmingly more important: no one would suggest that the variation in weights of a sample of 100 rats was caused by inaccurate weighing; instead, one would argue that the rats were not all identical and that their weights deviated, not from their 'correct' values, but from some idealized 'population' value. It is open to argument whether enzyme molecules vary in this way and, even if they do, it is questionable whether one would expect 'samples' from a stock solution, each containing many billions of molecules, to vary significantly. Nonetheless, the possibility should not be rejected out of hand; in particular, Reich has argued that random variations in enzyme kinetic experiments are too large to be explained solely by poor technique and instrumentation.

Systematic variations are also of two types, namely those that result from lack of fit, and those that result from errors that affect all observations in a systematic way. Lack of fit is the type of variation that results from fitting observations to the wrong equation. If one made perfectly accurate observations for an enzyme that obeyed the Michaelis–Menten equation, but fitted them to the equation for first-order kinetics, the best possible line would not fit all of the observations. This type of error can often be identified by statistical tests (*see* Section 10.8) and, provided that the correct equation can be found, it can be eliminated.

The second type of systematic variation is more serious, because it is present in all experiments, and cannot readily be detected. Any errors that affect the entire experiment, such as incorrect estimates of the enzyme and substrate concentrations in stock solutions, cause errors in the fitted constants, but they do not increase the estimated statistical error because they do not cause deviations from the best-fit equation. The only way they can be detected is by repeating the entire experiment, including all preliminary preparations. One generally finds in this case that day-to-day variations in estimated constants are much greater than the statistical errors estimated from within-experiment variation.

The practical importance of systematic errors is that the numerical precision of a kinetic constant should not be taken too seriously, but this does not mean that statistical calculations are of no value, because comparisons of numbers within an experiment can be made legitimately. For example, suppose that a suspected inhibitor of an enzyme raised the estimated value of K_m^{app} by an amount that was less than the day-to-day variation but significantly more than the within-experiment variation as estimated from statistical calculations. Provided that the result was reproducible and the inhibited and uninhibited rates were measured in a single experiment with the same stock solutions, it would indicate a significant degree of inhibition.

10.2 Variance

In statistics, one must always make a distinction between the *true value* of a quantity, which is an unknown constant and is generally denoted by a Greek letter, e.g. β, and an *estimate* of that quantity, which is a variable (as we can assign any value we like to it) and is generally denoted by the corresponding Roman letter, e.g. b. One may also consider the *best estimate*, by some criterion, denoted by $\hat{b}$ (pronounced 'b-hat'), which again is a constant, as there will normally be only one value of $\hat{b}$ that satisfies any defined criterion. (This vagueness about criteria is deliberate; it is a mistake to assume that there is a unique criterion for assessing closeness of fit that can be applied rigidly to all experiments.) Different symbols should be used for b and β because, although we hope that they will be equal, they are never exactly equal in practice. It is very important in statistical studies to be clear which quantities are variables and which are constants, and it should be noted that these do not in general correspond with what might be expected. Thus, the unknown constants α and β in an equation $y = \alpha + \beta x$ must be replaced with the known variables a and b if we wish to estimate them; but the quantities x and y, which are variables to the experimenter, are constants to the statistician, because they cannot be altered during statistical analysis.

As any estimate b differs from the true value β, it is natural to demand some estimate of the magnitude of the error, $b - \beta$. Analysis of the way in which different b values differ from one another provides information about the way in which they differ individually from β. If we determine a large number, n, of values, b_i, it may happen that the mean value, $\frac{1}{n} \sum b_i$, approaches β as n approaches infinity. If this is true, then b is, by definition, an unbiassed estimator* of β; in the remainder of this section, it is assumed that we are discussing unbiassed estimators, although estimators in the physical sciences are never truly unbiassed because it is never possible to allow for all sources of systematic error. Hence, for an unbiassed estimator, the mean error, i.e. $\frac{1}{n} \sum (b_i - \beta)$, approaches zero as n approaches infinity, and it does not provide a useful measure of the variability of b. However, the mean *squared* error, $\frac{1}{n} \sum (b_i - \beta)^2$, does not approach zero as n increases because every term in the summation is positive. Instead, it approaches a definite limit, known as the *variance* of b_i, or $\sigma^2(b_i)$, which does provide a measure of the variability of b_i.

The variance of b_i is defined as

---

* Not all statistical authorities make a distinction between an *estimate*, which is a particular value in a particular context, and an *estimator*, which defines a general class of estimates, but the distinction is sometimes useful for clarity and will be made in this account.

$$\sigma^2(b_i) = \lim_{n \to \infty} \frac{1}{n} \sum (b_i - \beta)^2 \qquad (10.1)$$

but it cannot be measured, because β is unknown and because one cannot carry out an infinite number of determinations. However, it can be estimated from the sample variance, $s^2(b_i) = \frac{1}{n} \sum (b_i - \hat{b})^2$, where n is finite and $\hat{b}$ is the best-fit estimate of β, by some criterion. A simple criterion for defining $\hat{b}$ is to define it as the value that makes $s^2(b_i)$ a minimum. Then we have

$$\frac{ds^2(b_i)}{d\hat{b}} = -\frac{2}{n} \sum (b_i - \hat{b}) = 0$$

i.e. $\sum \hat{b} = \sum b_i$. As $\hat{b}$ is the same for each i, $\sum \hat{b}$ is simply $n\hat{b}$, and so $\hat{b} = \frac{1}{n} \sum b_i$, i.e. the mean of all the b_i.

In general, the sample variance, $s^2(b_i)$, underestimates the true variance $\sigma^2(b_i)$ because, of all the ways in which $\hat{b}$ might be defined, we normally choose the definition that makes $s^2(b_i)$ a minimum. Hence $s^2(b_i)$ is a biassed estimator of $\sigma^2(b_i)$. While it is intuitively obvious that a correction for this bias is needed, the magnitude of the correction is not obvious, but a derivation of its magnitude would be beyond the scope of this book. We therefore state, without proof, that the bias can be corrected by multiplying the sample variance by $n/(n-1)$. In the more general case, where p parameters may have been estimated from n observations, the correction factor is $n/(n-p)$, i.e.

$$\sigma^2(b_i) \approx \left(\frac{n}{n-p}\right) s^2(b_i) \qquad (10.2)$$

When tabulating results, it is common practice to replace the variance of an estimate by the square root of the variance, which is known as the *standard error*. The practical advantage of this procedure is that the standard error of any value has the same dimensions as the value itself, so that one can write (for example) $\hat{b} - 3.21 \pm 0.12$, where the first number is the estimate of β and the second is the standard error of the estimate. Nonetheless, the variance is much more amenable to algebraic manipulation, as squares are simpler to deal with than square roots, and for that reason it is preferred in theoretical discussions.

In practice, we are more likely to be interested in the variance of the best estimate, $\hat{b}$, than in that of the individual values, b_i. In order to calculate the variance of $\hat{b}$, we must first determine the variance of a sum; so, consider the sum of two numbers, $(x + y)$, where x and y have variances defined by

$$\sigma^2(x) = \lim_{n \to \infty} \frac{1}{n} \sum (x_i - \mu_x)^2$$

$$\sigma^2(y) - \lim_{n \to \infty} \frac{1}{n} \sum (y_i - \mu_y)^2$$

where μ_x and μ_y are the population ('true') means of x_i and y_i, and we postulate that each value of x_i or y_i is drawn from an infinite population of possible

171

values. Then, it is logical to define the variance of $(x+y)$ as

$$\sigma^2(x+y) = \lim_{n\to\infty} \frac{1}{n}\sum(x_i+y_i-\mu_x-\mu_y)^2$$

Each term in the summation can be analysed as

$$(x_i+y_i-\mu_x-\mu_y)^2 = (x_i-\mu_x)^2+2(x_i-\mu_x)(y_i-\mu_y)+(y_i-\mu_y)^2$$

and so

$$\sigma^2(x+y) = \sigma^2(x)+2\,\text{cov}(x,y)+\sigma^2(y) \tag{10.3}$$

where $\text{cov}(x,y)$ is a new quantity, known as the *covariance* of x and y, and is defined as

$$\text{cov}(x,y) = \lim_{n\to\infty} \frac{1}{n}\sum(x_i-\mu_x)(y_i-\mu_y) \tag{10.4}$$

Clearly, the covariance is a measure of the tendency of x and y to vary in unison: if there is any systematic source of variation that affects both x and y, the covariance is significant. It is positive if x and y are affected in the same direction, and negative if they are affected in opposite directions. In the absence of systematic error, the covariance is zero and equation 10.3 simplifies to

$$\sigma^2(x+y) = \sigma^2(x)+\sigma^2(y) \tag{10.5}$$

Applying this result to the definition of $\hat{b}$ as $\frac{1}{n}\sum b_i$, we obtain

$$\sigma^2(\hat{b}) = \frac{1}{n^2}\sigma^2(\sum b_i) = \frac{1}{n^2}\sigma^2(b_1+b_2+\ \cdots\ +b_n)$$

$$= \frac{1}{n^2}\left[\sigma^2(b_1)+\sigma^2(b_2)+\ \cdots\ +\sigma^2(b_n)\right]$$

and if every b_i has the same variance, we obtain

$$\sigma^2(\hat{b}) = \frac{1}{n^2}n\ \sigma^2(b_i) = \frac{1}{n}\sigma^2(b_i) \tag{10.6}$$

Sometimes one requires the variance of a product. Provided that the individual variances are small, one can calculate the variance of a product xy fairly easily, as follows. If $x = \mu_x+\varepsilon_x$ and $y = \mu_y+\varepsilon_y$, then

$$xy = (\mu_x+\varepsilon_x)(\mu_y+\varepsilon_y) = \mu_x\mu_y+\mu_y\varepsilon_x+\mu_x\varepsilon_y+\varepsilon_x\varepsilon_y$$

Provided that it can be assumed that $\varepsilon_x \ll \mu_x$ and $\varepsilon_y \ll \mu_y$, then $\varepsilon_x\varepsilon_y$ is negligible and its variance is also negligible. Further, as μ_x and μ_y are constants, the variance of $\mu_x\mu_y$ is zero. Therefore, in calculating the variance of xy, we need only consider the two middle terms of the expansion and we can apply equation 10.3 for the variance of a sum, to give

$$\sigma^2(xy) = \mu_y^2\sigma^2(x)+2\mu_x\mu_y\,\text{cov}(x,y)+\mu_x^2\sigma^2(y) \tag{10.7}$$

As μ_x and μ_y are usually unknown, in practice they are replaced with estimates, $\hat{x}$ and $\hat{y}$.

The variance of a quotient can be derived in a similar way, and is given by

$$\sigma^2(x/y) = \frac{\sigma^2(x)}{\mu_y^2} - \frac{2\mu_x \text{cov}(x, y)}{\mu_y^3} + \frac{\mu_x^2 \sigma^2(y)}{\mu_y^4} \tag{10.8}$$

Again, this equation is only an approximation and is valid for small errors, and again the unknowns are normally replaced with estimates in practice.

Finally, it is useful to know the variance of a reciprocal. This is particularly valuable in enzyme kinetics because of the insight that it provides into the inadequacies of the double-reciprocal plot (Section 10.5). The error in $1/x$ is

$$\frac{1}{x} - \frac{1}{\mu} = (\mu - x)/x\mu = -\varepsilon/x\mu$$

where μ is the true value of x and ε is the error in x. Hence the variance of $1/x$ is given by

$$\sigma^2(1/x) = \sigma^2(x)/x^2\mu^2 \approx \sigma^2(x)/x^2\hat{x}^2 \approx \sigma^2(x)/x^4 \tag{10.9}$$

the last form being appropriate only when x is the best available estimate of μ.

10.3 Simple linear regression

At first sight, the term *regression* implies going back, and the relevance of this meaning to the use of the word in statistics is far from obvious. In fact, when we fit a calculated line to a set of experimental observations, we hope that we are 'going back' to the physical reality that gave rise to the observations. For example, if we have a set of n values of a variable, y, measured at a series of values of another variable, x, and we assume that the values of y ought to obey the equation for a straight line:

$$y_i = \alpha + \beta x_i$$

where the subscripts i indicate that the equation refers specifically to the ith observation, then we may attempt to 'go back' to the values of the constants α and β that gave rise to the values of y that we observed, and this process is called regression.

In all real experiments, observations are subject to error and so we cannot regard each measured y_i as an exact measure of $\alpha + \beta x_i$ in the above case; instead, we must write

$$y_i = \alpha + \beta x_i + \varepsilon_i \tag{10.10}$$

where ε_i is the error in y_i (note that x_i is assumed to be error-free). Unfortunately, the ε_i values are unknown, and they remain unknown no matter how many observations we make and no matter how much analysis we carry out, because there must always be two more unknowns than equations. Hence there can never be sufficient information to calculate α and β. All that is

173

possible is to estimate the most likely values of α and β, on the basis of some assumptions about the nature of the ε_i values.

First, equation 10.10 must be re-written in terms of known quantities:

$$y_i = a + bx_i + e_i \qquad (10.11)$$

where a, b and e_i are *assumed* quantities that may approximate the true values of α, β and ε_i. In order to avoid making expressions unnecessarily complex, subscripts i will be omitted in the remainder of this chapter, except where they are required for clarity. Similarly, for all summations the sign $\sum$ will indicate summation over all observations, i.e. from $i = 1$ to n. The e values are called deviations (or residuals) rather than errors, because they need not be equal to the true errors, ε. As explained in the previous section, this apparently pedantic distinction between true and estimated quantities is essential for clarity in discussing the theory of regression. Provided that the experiment is unbiassed, i.e. that each ε can be assumed to come from a distribution with zero mean, then it is natural to assume that α and β will best be approximated by values of a and b that make the total deviation as small as possible. In order to achieve this, it is first necessary to define the (deliberately) vague term 'total deviation' more precisely. The most obvious definition is simply $\sum e$, but this is ruled out by the fact that it does not define a and b uniquely, as for *any* value of a there is a value of b that makes $\sum e$ zero, because the summation includes both positive and negative terms, which can be made to cancel exactly without in fact achieving a close fit to the observations. This difficulty can be removed either by neglecting the signs and minimizing $\sum |e|$, the sum of errors with the signs omitted, or by minimizing $\sum e^2$, the sum of squared errors. There is little fundamental objection to the use of $\sum |e|$ as a measure of closeness of fit, but it is ruled out in most practical applications by the fact that it leads to hopelessly difficult algebra in all but the simplest cases. In practice, therefore, we minimize $\sum e^2$, which is called the *sum of squares*. (In many elementary accounts, the sum of squares is introduced in a more enthusiastic way, and is claimed to possess certain fundamental advantages over other criteria. However, close examination shows that, while the logic may be faultless, the premises upon which it is built are not, and generally include several palpable falsehoods about the nature of experimental error. It is therefore safer and more honest to admit that the main virtue of the sum of squares is its convenience in algebraic manipulation.)

It may happen that we know, *a priori*, that some observations are more precise than others. In such a case, it is logical to give more weight to the better observations in assessing the closeness of fit. Therefore, rather than the unweighted sum of squares, $\sum e^2$, it is usually preferable to minimize a *weighted* sum of squares, $SS = \sum we^2$, where each w value is a weighting factor. This, of course, assumes some way of knowing what the weights ought to be. In finding the weighted mean of a set of values, it is not difficult to show that the variance of the mean is a minimum if each value has a weight that is inversely proportional to its variance; this also applies in the more general problem of fitting observations to an equation. [This conclusion requires the assumption that the errors in the observations are uncorrelated, i.e. that $\text{cov}(\varepsilon_i, \varepsilon_j) = 0$

for all $i \neq j$. This assumption is rarely likely to be exactly true, but the correct weights are very difficult to calculate if it is not made.] In practice, one is unlikely to know the exact variances of the observations, but it may often be possible to make a plausible guess about the way experimental error varies with the quantities being measured. In enzyme kinetics, it is *never* appropriate to carry out an unweighted fit to a straight line, and weights should therefore always be included in the definition of the sum of squares.

Rearranging equation 10.11, we obtain

$$e = y - a - bx$$

and the sum of squares is given by

$$SS = \sum we^2 = \sum w(y - a - bx)^2$$

For any value of b, SS varies with a according to a quadratic equation, so that a plot of SS against a is a parabola with a slope at any point given by

$$\frac{\partial SS}{\partial a} = -2 \sum w(y - a - bx) = -2 \sum wy + 2a \sum w + 2b \sum wx$$

and similarly, for any value of a, a plot of SS against b has a slope given by

$$\frac{\partial SS}{\partial b} = -2 \sum wxy + 2a \sum wx + 2b \sum wx^2$$

For SS to be a minimum, both slopes must simultaneously be zero. Therefore, if $\hat{a}$ and $\hat{b}$ are defined as the values of a and b that make SS a minimum, we have

$$\left. \begin{array}{l} \hat{a} \sum w + \hat{b} \sum wx = \sum wy \\ \hat{a} \sum wx + \hat{b} \sum wx^2 = \sum wxy \end{array} \right\} \qquad (10.12)$$

These are a pair of ordinary simultaneous equations (sometimes known as the *normal equations*) that can readily be solved for the unknowns $\hat{a}$ and $\hat{b}$:

$$\hat{b} = \frac{\sum w \sum wxy - \sum wx \sum wy}{\sum w \sum wx^2 - (\sum wx)^2} \qquad (10.13)$$

$$\hat{a} = \frac{\sum wy - \hat{b} \sum wx}{\sum w} \qquad (10.14)$$

These equations are convenient as written for calculating $\hat{a}$ and $\hat{b}$, but they do not show directly how the variances of $\hat{a}$ and $\hat{b}$ can be found. However, with a little ingenuity, equation 10.13 can be rearranged to read

$$\hat{b} = \frac{\sum (x - \bar{x})wy}{\sum (x - \bar{x})wx} \qquad (10.15)$$

where $\bar{x} = \sum wx / \sum w$ is the weighted mean of the x values. This equation is of the form

$$\hat{b} = \sum uy \qquad (10.16)$$

where each u value is a constant given by

$$u = \frac{(x - \bar{x})w}{\sum(x - \bar{x})wx} \tag{10.17}$$

As the errors in the y values are, by the original hypothesis, uncorrelated, the variance of $\hat{b}$ is found by applying a generalized form of equation 10.5 to equation 10.16, i.e.

$$\sigma^2(\hat{b}) = \sum u^2 \sigma^2(y) \tag{10.18}$$

Now, each weight w is, by definition, inversely proportional to $\sigma^2(y)$ and so, for each i, we can write $\sigma^2(y_i) = \sigma^2_{exp}/w_i$, where σ^2_{exp} is a constant independent of i, and is known as the *experimental variance*. Equation 10.18 can therefore be written as

$$\sigma^2(\hat{b}) = \sigma^2_{exp} \sum u^2/w \tag{10.19}$$

and, after some tedious but not difficult algebra, this can be rearranged into the form

$$\sigma^2(\hat{b}) = \frac{\sigma^2_{exp} \sum w}{\sum w \sum wx^2 - (\sum wx)^2} \tag{10.20}$$

Although this expression appears more complex than equation 10.19, it is more convenient to apply because it contains only summations that appear also in equation 10.13; thus it requires negligible computation beyond that required for evaluating $\hat{b}$.

Further expressions can be derived in a similar manner for the variance of $\hat{a}$ and the covariance of $\hat{a}$ and $\hat{b}$:

$$\sigma^2(\hat{a}) = \frac{\sigma^2_{exp} \sum wx^2}{\sum w \sum wx^2 - (\sum wx)^2} \tag{10.21}$$

$$\text{cov}(\hat{a}, \hat{b}) = \frac{-\sigma^2_{exp} \sum wx}{\sum w \sum wx^2 - (\sum wx)^2} \tag{10.22}$$

All of these equations contain σ^2_{exp}, which is unknown, but can readily be estimated from the sum of squares, i.e. $\sigma^2_{exp} \approx SS/(n-2)$. Division by $(n-2)$ rather than n corrects for the bias that derives from the fact that SS is a minimum value, and must therefore be less than the true value of $\sum w\varepsilon^2$ (cf. equation 10.2).

It is worthy of comment that none of equations 10.13–10.22 is symmetrical in x and y and so regression of x on y instead of y on x would give a different best-fit straight line. This asymmetry is present in the initial assumption that y is subject to error but x is not, so that deviations from the line are measured parallel with the y-axis and not, as one might suppose, at right-angles to the line. Measurement of errors at right-angles to the line might seem to avoid the need for an assumption about which variable is subject to error, but in fact it creates far worse difficulties than it solves: not only is the algebra much more difficult but, in addition, the best-fit line varies according to the units in which x and y are measured because the lengths of lines drawn on a graph have meaningful dimensions only if they are parallel with one or other axis.

The straight line $y = \alpha + \beta x + \varepsilon$ is the simplest of a general class of models known as *linear models*. The word 'linear' in this context is not simply a trivial repetition of the fact that a straight line is a straight line; it means that the model is of the form

$$y_i = \beta_0 + \beta_1 x_{1i} + \beta_2 x_{2i} + \cdots + \varepsilon_i$$

where y_i is the only variable subject to error, $\beta_0, \beta_1, \beta_2, \ldots$, are the parameters of the model, and $x_{1i}, x_{2i}, \ldots$, are variables that are known exactly. In other words, a linear model is a linear function of the parameters. It may also be linear in the observations, but this is irrelevant. Hence the equation $y = \alpha + \beta x + \gamma x^2 + \varepsilon$ is linear, even though it describes a curve and not a straight line, whereas the equation $x/\alpha + y/\beta = 1 + \varepsilon$ is non-linear, even though it defines a straight line. The importance of linearity is that linear models can be analysed very simply, because they give rise to equations similar to equations 10.12 that can be solved exactly in a single step. Unfortunately, many of the models that occur in enzyme kinetics (and many other physical sciences) are non-linear, and thus require special techniques for their analysis. The practical value of linear regression is that in many cases models can be expressed in such a way that linear regression can be used in their analysis. Several examples of this are discussed in this chapter.

10.4 Fitting the Michaelis–Menten equation

Just as the idealized equation for a straight line, $y = \alpha + \beta x$, must be modified so as to include experimental error, $y = \alpha + \beta x + \varepsilon$, if it is to represent a real experimental situation, so also is the Michaelis–Menten equation, $v = \mathscr{V} s/(\mathscr{K}_m + s)$, incomplete until error is introduced:

$$v = \frac{\mathscr{V} s}{\mathscr{K}_m + s} + \delta \tag{10.23}$$

(In the absence of suitable Greek letters, the symbols $\mathscr{V}$ and $\mathscr{K}_m$ represent the true values of V and K_m. In addition, the symbols δ, d and u will be used instead of ε, e and w, because we shall require the original symbols for referring back to the results of Section 10.3.) This simple change immediately indicates the fault in the double-reciprocal plot and other linear transformations of the Michaelis–Menten equation discussed in Section 2.5: it is not easy to see at first how a perfectly valid algebraic transformation can give an invalid result, but once it is realized that it is not the transformation that is invalid but the starting point, the problem disappears.

In order to fit data to equation 10.23, we can minimize the weighted sum of squares, $SS = \sum u d^2$, where d is an approximation to δ obtained by replacing $\mathscr{V}$ and $\mathscr{K}_m$ with estimates $\hat{V}$ and $\hat{K}_m$, respectively. In principle, therefore, the problem is exactly the same as that discussed in the previous section, but practical difficulties arise because equation 10.23 is non-linear and so SS cannot be minimized by simple direct means. Nonetheless, there are several indirect means of minimizing SS, of which we shall describe one, based mainly

177

on the work of Johansen and Lumry (1961) and of Wilkinson (1961).

If an error term is introduced into the Michaelis–Menten equation *after* transformation to a linear form, instead of (correctly) before:

$$\frac{s}{v} = \frac{\mathscr{K}_m}{\mathscr{V}} + \frac{1}{\mathscr{V}} s + \varepsilon$$

then the result is not a true transformation of equation 10.23 and ε is not equal to δ (which is why we need a different symbol). In fact we can show, by a simple piece of algebra, that

$$\delta = \frac{-\mathscr{V} v \varepsilon}{\mathscr{K}_m + s} \tag{10.24}$$

As $\mathscr{V}$, $\mathscr{K}_m$, δ and ε are all unknown, we must replace them with estimates V, K_m, d and e, respectively, and equation 10.24 becomes

$$d = \frac{-V v e}{K_m + s} \tag{10.25}$$

For $V \approx \mathscr{V}$, $K_m \approx \mathscr{K}_m$ and $v \approx \mathscr{V} s/(\mathscr{K}_m + s)$, i.e. in the vicinity of the best-fit solution for data that are worth fitting, $V/(K_m + s) \approx v/s$. Thus

$$d \approx \frac{-v^2 e}{s} \tag{10.26}$$

These results can be used in order to minimize SS in a straightforward way, provided that the weights u can be defined. As discussed in the previous section, the weights should properly reflect the way in which the variance of v varies with v. In practice, this is usually unknown, but the truth probably lies between two limiting hypotheses: (*i*) *simple errors in v*, where each velocity has the same standard error, and so $u = 1$ for every observation; and (*ii*) *relative errors in v*, where each velocity has a standard error proportional to its true value $\mathscr{V} s/(\mathscr{K}_m + s)$, and so $u = (\mathscr{K}_m + s)^2/\mathscr{V}^2 s^2$ for each observation.

In the first case, the minimization of SS is simply the minimization of

$$SS = \sum d^2 = \sum \frac{V^2 v^2 e^2}{(K_m + s)^2} \approx \sum \frac{v^4 e^2}{s^2} \tag{10.27}$$

The approximate form has the advantage that it contains no unknowns; so, as a first step in minimizing SS, we can minimize $\sum v^4 e^2/s^2$ by means of the linear regression formulae given in the previous section. Thus, substituting $\hat{a} = \hat{K}_m/\hat{V}$, $\hat{b} = 1/\hat{V}$, $x = s$, $y = s/v$ and $w = v^4/s^2$ into equations 10.13 and 10.14 and rearranging, we obtain

$$\hat{V} \approx \frac{\sum v^4/s^2 \sum v^4 - (\sum v^4/s)^2}{\sum v^4/s^2 \sum v^3 - \sum v^4/s \sum v^3/s} \tag{10.28}$$

$$\hat{K}_m \approx \frac{\sum v^4 \sum v^3/s - \sum v^4/s \sum v^3}{\sum v^4/s^2 \sum v^3 - \sum v^4/s \sum v^3/s} \tag{10.29}$$

For many purposes, it is adequate to stop at this point; indeed, it may be argued that any attempt to refine these values is supererogatory as it implies

178

an unwarranted degree of confidence in the weighting scheme. However, for completeness we shall show how to minimize SS exactly. Once approximate values of $\hat{V}$ and $\hat{K}_m$ are available, the first-approximation weights, v^4/s^2, can be replaced with refined weights, $V_0^2 v^2/(K_0+s)^2$, where V_0 and K_0 are the approximate values of $\hat{V}$ and $\hat{K}_m$ found from equations 10.28 and 10.29. Then the refined values of $\hat{V}$ and $\hat{K}_m$ are given by

$$\hat{V} \approx \frac{\sum \dfrac{v^2}{(K_0+s)^2} \sum \dfrac{s^2v^2}{(K_0+s)^2} - \left(\sum \dfrac{sv^2}{(K_0+s)^2}\right)^2}{\sum \dfrac{v^2}{(K_0+s)^2} \sum \dfrac{s^2v}{(K_0+s)^2} - \sum \dfrac{sv^2}{(K_0+s)^2} \sum \dfrac{sv}{(K_0+s)^2}} \quad (10.30)$$

$$\hat{K}_m \approx \frac{\sum \dfrac{s^2v^2}{(K_0+s)^2} \sum \dfrac{sv}{(K_0+s)^2} - \sum \dfrac{sv^2}{(K_0+s)^2} \sum \dfrac{s^2v}{(K_0+s)^2}}{\sum \dfrac{v^2}{(K_0+s)^2} \sum \dfrac{s^2v}{(K_0+s)^2} - \sum \dfrac{sv^2}{(K_0+s)^2} \sum \dfrac{sv}{(K_0+s)^2}} \quad (10.31)$$

This process may be repeated, substituting the new value of $\hat{K}_m$ for K_0, until $\hat{K}_m$ does not change appreciably from one approximation to the next; in practice, this usually happens at about the fourth approximation. Notice that V_0 does not in fact appear in equations 10.30 and 10.31, as it cancels; consequently, it is unnecessary to evaluate $\hat{V}$ until the last step.

If we now examine the second weighting hypothesis, where each velocity is assumed to have a standard error that is proportional to its true value, we find, surprisingly, that the problem is much easier to solve. In this case, the proper weights, u, are given by $u = (\hat{K}_m+s)^2/\hat{V}^2 s^2$, where the best-fit estimates, $\hat{K}_m$ and $\hat{V}$, replace the unknowns $\mathcal{K}_m$ and $\mathcal{V}$. Hence the sum of squares is given by

$$SS = \Sigma u d^2 = \sum \frac{(\hat{K}_m+s)^2 V^2 v^2 e^2}{\hat{V}^2 (K_m+s)^2 s^2} \quad (10.32)$$

At the minimum, $K_m = \hat{K}_m$ and $V = \hat{V}$, by definition, and so

$$SS = \Sigma v^2 e^2/s^2 \quad (10.33)$$

This expression is exact only at the minimum itself but, as we are concerned with the minimum itself, that does not matter. As only known quantities are involved, SS can be minimized in a single step; substituting $\hat{a} = \hat{K}_m/\hat{V}$, $\hat{b} = 1/\hat{V}$, $x = s$, $y = s/v$ and $w = v^2/s^2$ into equations 10.13 and 10.14 and rearranging, we have the exact solutions:

$$\hat{V} = \frac{\Sigma v^2/s^2 \, \Sigma v^2 - (\Sigma v^2/s)^2}{\Sigma v^2/s^2 \, \Sigma v - \Sigma v^2/s \, \Sigma v/s} \quad (10.34)$$

$$\hat{K}_m = \frac{\Sigma v^2 \, \Sigma v/s - \Sigma v^2/s \, \Sigma v}{\Sigma v^2/s^2 \, \Sigma v - \Sigma v^2/s \, \Sigma v/s} \quad (10.35)$$

As this result (from Johansen and Lumry, 1961) is so much simpler to apply than the earlier one, it may be wondered why it is so rarely used. One answer is that it is not, or should not be, simply a matter of choice: the proper weights are not determined by analytical convenience but by the nature of the experimental errors. However, this is an unrealistic answer because it assumes that the nature of the experimental error has been investigated, something that has almost never been done. A more truthful answer is that in straight-line regression it is much more convenient to work with simple errors, and one therefore tends to apply wishful thinking to the problem and to regard simple errors as 'correct.' In enzyme kinetics, it is actually much more convenient to adopt the alternative hypothesis, albeit with as little justification. In practice, one can gain an idea about the correct weighting scheme by plotting d against v and d/v against v, after fitting the data with the weights that one wishes to test: if the errors are truly simple (or, to use the proper statistical term, if the d values are *homoscedastic*) the points should be scattered in a parallel band about the v axis when d is plotted, and in a cuspate band when d/v is plotted, as shown in the upper part of *Figure 10.1*.

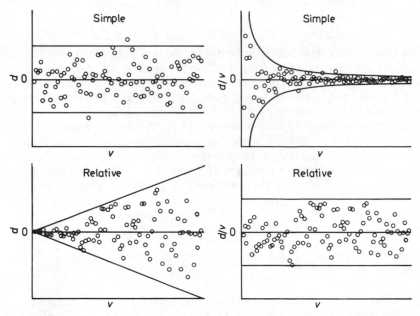

Figure 10.1 Scatter plots for assessing the correctness of a weighting scheme: The upper pair of plots show the expected results of plotting the deviation from the fitted line, d, against the velocity, v, and of plotting d/v against v, in the event that all the velocities have the same standard deviation ('simple errors'). The lower pair of plots show the expected results in the event that the standard deviation of each velocity is proportional to its true value ('relative errors'). In each plot, limits are drawn for deviations of twice the standard deviation

If the errors are truly relative, however, so that d/v is homoscedastic rather than d, the points should be scattered in a wedge-shaped band when d is plotted, and in a parallel band when d/v is plotted, as shown in the lower part of *Figure 10.1*. The results shown are idealized, and in practice it is generally necessary to have a large number of points (50 or more) for the scatter to be clearly defined. In order to overcome this difficulty, one can combine the
180

results of several experiments, provided they were carried out under similar conditions by the same operator.

Sometimes one may feel that the two cases discussed are too extreme and that the best choice would be a compromise between them, with weights given by $u = (\hat{K}_m + s)/\hat{V}s$. In this case, the data can be fitted by an obvious adaptation of the method described for uniform weights, as the proper expression for the sum of squares is now

$$SS = \sum \frac{\hat{V}v^2 e^2}{\hat{K}_m + s} \approx \sum v^3 e^2/s^2 \tag{10.36}$$

which can be compared with equation 10.27 and the ensuing discussion. In the absence of any real information about the distribution of errors (i.e. usually), this may be a safer choice than either of the previous choices.

Figure 10.2 illustrates how the choice of weighting scheme affects the

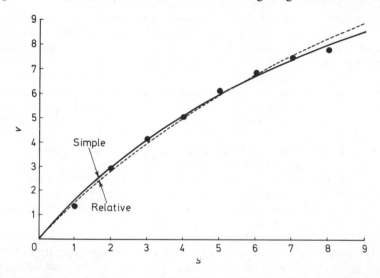

Figure 10.2 *Effect of weighting on estimates of kinetic parameters: The points shown were fitted to the Michaelis–Menten equation, giving $\hat{V} = 18.41$ and $\hat{K}_m = 10.41$ (solid line) when simple errors were assumed, and $\hat{V} = 23.61$ and $\hat{K}_m = 14.87$ (broken line) when relative errors were assumed. The difference between weighting schemes is most pronounced if the data are poor and extend over an inadequate range of s values (as here), or if the wrong equation is fitted, e.g. if significant substrate inhibition is not noticed*

definition of the 'best' values of V and K_m in practice. Notice particularly that although the two curves are similar the constants that define them are in poor agreement.

10.5 Final comments at the double-reciprocal plot

We can approach the results of the last section from a rather different point of view, by considering the problem as one of carrying out a least-squares fit to a straight line. In Section 10.3, it was seen that in such a case each observation (x, y) should be given a weight that is inversely proportional to the variance of the dependent variable, y. For the double-reciprocal plot, the dependent variable is $1/v$, which has a variance $\sigma^2(v)/v^2\hat{v}^2$, where $\hat{v}$ is the calculated value of v, i.e. $\hat{v} = \hat{V}s/(\hat{K}_m + s)$. Therefore, if all velocities have the same variance (an assumption that may not always be true, as discussed above), the proper weights in the double-reciprocal plot are proportional to $v^2\hat{v}^2$. As the calculated velocities are initially unknown, we can use weights v^4 calculated from the observed velocities as a first approximation. Then we can refine the result with weights $v^2\hat{v}^2$ until the results are self-consistent, i.e. until the estimate of $\hat{K}_m$ does not change from one approximation to the next. [Some workers, such as Cleland (1967), suggest the use of $\hat{v}^4$ weights for the second and subsequent approximations, but this suggestion is based upon an incorrect form of equation 10.9 for the variance of a reciprocal, and gives no improvement upon the first approximation.]

If the double-reciprocal plot is analysed with weights $v^2\hat{v}^2$, the calculations and final results are identical with those of the previous section (equations 10.28–10.31). Analogous analysis of the plot of s/v against s also gives identical results. We can then reasonably enquire in what sense the double-reciprocal plot is a 'worse' plot than the plot of s/v against s. The point is that for even a modest range of s values, the range of weights required for the double-reciprocal plot is enormous: e.g. if s varies from $0.2K_m$ to $2K_m$, with v varying from $0.167V$ to $0.667V$, the weights cover a 256-fold range. It is plainly impossible to assign one point 256 times more weight than another when looking at points on a plot; hence a double-reciprocal plot provides almost no useful information about the experimental error. Contrast this with the situation for the plot of s/v against s: here the proper weights are proportional to $v^2\hat{v}^2/s^2$, or approximately to v^4/s^2, and vary over a mere 3.24-fold range for the same range of v. In this case, visual examination of the plot gives a reasonably accurate account of the experimental error.

In this section we have hitherto assumed simple errors in v, i.e. all velocities have the same variance. If we now consider the alternative hypothesis of relative errors in v, the double-reciprocal plot remains the least satisfactory but the difference is far less marked: the proper weights for $1/v$ are now proportional to v^2 and vary over only a 16-fold range for the example considered above, whereas the proper weights for s/v are proportional to v^2/s^2 and vary over a 6.25-fold range. Incidentally, whichever plot is used in this instance, the proper weights should be calculated from the observed velocities, and 'refining' the result by the use of calculated velocities (Cleland, 1967) is a progression from an easy, exact solution to a laborious approximation.

10.6 Standard errors of $\hat{V}$ and $\hat{K}_m$

Expressions for the variances of $\hat{K}_m/\hat{V}$ and $1/\hat{V}$ can be obtained very simply by substituting $\hat{a} = \hat{K}_m/\hat{V}$, $\hat{b} = 1/\hat{V}$ and $x = s$ into equations 10.20 and 10.21:

$$\sigma^2(1/\hat{V}) = \frac{\sigma^2_{exp} \sum w}{\sum w \sum ws^2 - (\sum ws)^2} \tag{10.37}$$

$$\sigma^2(\hat{K}_m/\hat{V}) = \frac{\sigma^2_{exp} \sum ws^2}{\sum w \sum ws^2 - (\sum ws)^2} \tag{10.38}$$

$$\text{cov}(1/\hat{V}, \hat{K}_m/\hat{V}) = \frac{-\sigma^2_{exp} \sum ws}{\sum w \sum ws^2 - (\sum ws)^2} \tag{10.39}$$

The equations apply equally well to any weighting scheme. The variance of $\hat{V}$ is found from equation 10.37 by applying equation 10.9, the formula for the variance of a reciprocal:

$$\sigma^2(\hat{V}) \approx \frac{\hat{V}^4 \sigma^2_{exp} \sum w}{\sum w \sum ws^2 - (\sum ws)^2} \tag{10.40}$$

The variance of $\hat{K}_m$ is obtained from equations 10.37–10.39 by applying equation 10.8, the formula for the variance of a quotient:

$$\sigma^2(\hat{K}_m) \approx \hat{V}^2 \sigma^2_{exp} \left[\frac{\sum ws^2 + 2\hat{K}_m \sum ws + \hat{K}_m^2 \sum w}{\sum w \sum ws^2 - (\sum ws)^2} \right] \tag{10.41}$$

In spite of its complex appearance, this expression is easy to evaluate because all of the summations in it are known from the prior determination of $\hat{V}$ and $\hat{K}_m$. In all of the equations, σ^2_{exp} is estimated as $SS/(n-2)$ (cf. Section 10.3), and in each case the standard error is the square root of the variance.

An important point that is not evident in practice from the determination of the standard errors of $\hat{V}$ and $\hat{K}_m$ is that the estimates of $\hat{V}$ and $\hat{K}_m$ are invariably highly positively correlated. Hence the actual error in $\hat{V}$ is highly dependent on that in $\hat{K}_m$, and *vice versa*. Consequently, one ought to consider the precision of the two parameters jointly rather than separately. Now, although it is possible in principle to calculate a joint confidence region as a constant-SS contour (*see*, for example, Colquhoun, 1971), it is not a practical proposition for routine use because the necessary calculations are very tedious and it is profitable to consider practical alternatives. The simplest possibility is to remember the qualitative fact that $\hat{V}$ and $\hat{K}_m$ are always highly correlated, and that a significant positive error in one is most unlikely to be accompanied by a significant negative error in the other. A second possibility is to examine the standard error of $\hat{K}_m/\hat{V}$, which is readily calculated from equation 10.38, as well as those of $\hat{V}$ and $\hat{K}_m$. The relative error in $\hat{K}_m/\hat{V}$ is invariably smaller than that in $\hat{K}_m$, and it is often smaller than that in $\hat{V}$. Hence it is worthwhile to consider whether putative errors in $\hat{K}_m$ and $\hat{V}$ are consistent with the standard error of $\hat{K}_m/\hat{V}$.

Finally, it must be remembered that calculations of standard errors make no allowance for sources of error that affect all points in unison. These sources may be much more serious than the sources of error that are considered, and there is no reason why they should produce correlated effects in $\hat{V}$ and $\hat{K}_m$. Thus it is often found that day-to-day variations in both parameters are much greater than the standard errors calculated for any one day. Consider, for example, the following series of $\hat{K}_m$ values and standard errors determined by Hwang, Chen and Burris (1973) for bacterial nitrogenase in five experiments: 0.131 ± 0.016, 0.059 ± 0.008, 0.102 ± 0.027, 0.09 ± 0.02, 0.12 ± 0.03 atm of nitrogen. The standard deviation between experiments for these results is 0.028 atm of nitrogen, so that there must be significant variation between experiments not taken into account in the individual standard error estimates. Moreover, the two extreme values have the two smallest standard error estimates, so that there seems to be very little relationship between the calculated errors and the actual errors in the five values. This type of result is by no means unusual (all that is unusual is the clarity with which it is presented), and it indicates that standard error estimates should be used with great caution. As $\hat{V}$ is directly affected by uncertainty in the true enzyme concentration, it is likely that results for $\hat{V}$ would show even less agreement between day-to-day variations and standard error estimates.

10.7 General linear model and applications to more complex cases

In order to fit data to equations more complex than the Michaelis–Menten equation, with three or more parameters, we must first consider a generalized straight line, the *general linear model*:

$$y_i = \beta_1 x_{1i} + \beta_2 x_{2i} + \cdots + \beta_p x_{pi} + \varepsilon_i \tag{10.42}$$

In this equation, each x requires two subscripts: the first defines the nature of x and the second the number of the observation; thus x_{1i} might be the ith substrate concentration, x_{2i} might be the ith inhibitor concentration, etc. Although no constant intercept appears explicitly in the equation, it is not excluded, because x_{1i} may be defined as 1 for each i, in which case β_1 would be a constant corresponding to α in Section 10.3. (In some texts, the constant is defined as β_0, but this leads to unnecessary confusion because then the number of parameters is $p+1$ rather than p.)

For any estimates of the parameters, $b_1, b_2, \ldots, b_p$, equation 10.42 can be written as

$$y_i = b_1 x_{1i} + b_2 x_{2i} + \cdots + b_p x_{pi} + e_i \tag{10.43}$$

and the weighted sum of squares can be defined as

$$SS = \sum w_i e_i^2 = \sum w_i (y_i - b_1 x_{1i} - b_2 x_{2i} - \cdots - b_p x_{pi})^2$$

Partially differentiating with respect to each parameter in turn, we obtain

$$\frac{\partial SS}{\partial b_1} = -2 \sum w_i x_{1i} y_i + 2b_1 \sum w_i x_{1i}^2 + 2b_2 \sum w_i x_{2i} x_{1i} + \cdots + 2b_p \sum w_i x_{pi} x_{1i}$$

$$\frac{\partial SS}{\partial b_2} = -2 \sum w_i x_{2i} y_i + 2b_1 \sum w_i x_{1i} x_{2i} + 2b_2 \sum w_i x_{2i}^2 + \cdots + 2b_p \sum w_i x_{pi} x_{2i}$$

etc. In order to find the estimates $\hat{b}_1, \hat{b}_2 \cdots \hat{b}_p$ that minimize SS, we must set all p expressions to zero, replace each b_j with $\hat{b}_j$ and rearrange:

$$\left.\begin{array}{l} \hat{b}_1 \sum w_i x_{1i}^2 + \hat{b}_2 \sum w_i x_{2i} x_{1i} + \cdots + \hat{b}_p \sum w_i x_{pi} x_{1i} = \sum w_i x_{1i} y_i \\ \hat{b}_1 \sum w_i x_{1i} x_{2i} + \hat{b}_2 \sum w_i x_{2i}^2 + \cdots + \hat{b}_p \sum w_i x_{pi} x_{2i} = \sum w_i x_{2i} y_i \end{array}\right\} \qquad (10.44)$$

etc. Instead of the two simultaneous equations, equations 10.12, for the straight-line case, we now have p simultaneous equations. Solution of these equations is not the elementary problem that it might appear to be, because of serious arithmetical difficulties that arise whenever there are more than a small number of simultaneous equations to be solved. These are considered in Section 10.8. We can formalize the problem by stating that we require a set of coefficients, c_{jk}, such that:

$$\hat{b}_1 = c_{11} \sum w_i x_{1i} y_i + c_{21} \sum w_i x_{2i} y_i + \cdots + c_{p1} \sum w_i x_{pi} y_i$$
$$\hat{b}_2 = c_{12} \sum w_i x_{1i} y_i + c_{22} \sum w_i x_{2i} y_i + \cdots + c_{p2} \sum w_i x_{pi} y_i$$

etc. The original set of equations, equations 10.44, are said to be *inverted* in the sense that the unknowns $\hat{b}_1, \hat{b}_2, \ldots, \hat{b}_p$ are now written explicitly, whereas the terms on the right-hand side in equation 10.44, $\sum w_i x_{1i} y_i$, $\sum w_i x_{2i} y_i, \ldots$ are now written as if they were unknowns. Thus the *matrix* of coefficients c_{jk} is said (by definition) to be the *inverse* of the original matrix of coefficients $\sum w_i x_{ji} x_{ki}$. As matrix-inversion routines are available in all modern computer libraries, we need not concern ourselves with the mechanics of calculating the coefficients c_{jk}. The important point is that not only do they provide a straightforward way of evaluating all of the $\hat{b}_j$ values, but they also provide a simple way of calculating all of the variances and covariances. For any parameter,

$$\sigma^2(\hat{b}_j) = c_{jj} \sigma_{exp}^2$$

and for any pair of parameters,

$$\mathrm{cov}(\hat{b}_j, \hat{b}_k) = c_{jk} \sigma_{exp}^2$$

where σ_{exp}^2 is estimated from the sum of squares in the usual way as $SS/(n-p)$. The validity of this result is not obvious and cannot easily be demonstrated without a long excursus into matrix algebra. A textbook on regression (e.g. Draper and Smith, 1966) should be consulted for more information.

Most of the equations commonly encountered in steady-state kinetics can be written in the form of the general linear model. Consider, for example, the equation for the initial rate of an irreversible reaction subject to product inhibition (equation 2.22):

$$v_i = \frac{V^f a_i}{K_m^A (1 + p_i/K_s^P) + a_i} + d_i \qquad (10.45)$$

and suppose that we wish to minimize the weighted sum of squares, $SS = \sum u_u d_i^2$. The equation can be written as

$$\frac{a_i}{v_i} = \frac{K_m^A}{V^f} + \frac{K_m^A}{V^f K_s^P} p_i + \frac{1}{V^f} a_i + e_i \qquad (10.46)$$

where e_i is not equal to d_i, but to $-d_i a_i/v_i^2$ approximately (cf. equation 10.26).

This corresponds to equation 10.43 with $y_i = a_i/v_i$, $b_1 = K_m^A/V^f$, $b_2 = K_m^A/V^f K_s^P$, $b_3 = 1/V^f$, $x_{1i} = 1$ (for all i), $x_{2i} = p_i$ and $x_{3i} = a_i$. The proper weights w_i are given by $w_i = u_i v_i^4/a_i^2$ in the first approximation, or $w_i = u_i v_i^2 \hat{v}_i^2/a_i^2$, where $\hat{v}_i$ is the velocity calculated with the best available estimates of the parameters. Exactly the same considerations apply to the definitions of the weights u_i as in the simpler case of the Michaelis–Menten equation (Section 10.4): if we assume simple errors in the velocities, we have $u_i = 1$ for each i; if we assume relative errors in the velocities, we have $u_i = 1/\hat{v}_i^2$. In the latter case, $\hat{v}_i^2$ cancels from the expression for w_i, and so $w_i = v_i^2/a_i^2 b_i^2$ is an exact expression and does not require refinement; again, this is the same as in the simpler case.

A high proportion of steady-state rate equations are of essentially the same form as equation 10.46, i.e. the right-hand side consists of a fraction with a single term in the numerator, and a linear expression in the denominator. All of these equations can be treated in the same way as equation 10.46. Equations with more than one term in the numerator, e.g. the equation for hyperbolic inhibition or activation (equation 4.7), are more difficult to fit and require more versatile non-linear regression techniques. Swann (1969) has discussed many of these techniques and Wharton *et al.* (1974) described and discussed a specific example in enzyme kinetics.

10.8 Some difficulties in fitting data

One could fill a separate book with a discussion of the practical difficulties that can arise in attempting to apply the methods described in this chapter, but some problems occur so frequently that some mention must be made of them.

Major difficulties often arise from a source that is unexpected to the in-experienced, namely rounding error. For most non-statistical purposes, it is sufficient to carry out calculations with one or two more significant figures than one expects to have in the final answer. However, this is not nearly enough for most statistical calculations, because these always involve measuring the difference between two large numbers at some stage, which always results in the loss of some significant digits. Consider, for example, the difference $1.38204 - 1.38195 = 0.00009$; although there are six significant digits in each of the original numbers, there is only one in the difference. It is instructive to examine the calculation of $\hat{K}_m$, by means of equation 10.28, for the sample set of data shown in *Table 10.1*. As the velocities are expressed to only two decimal places and the formula provides only a first approximation for $\hat{K}_m$, one might easily assume that it would be sufficient to carry only three decimal places at each step in the calculation. However, as shown, this gives a final answer of $\hat{K}_m = 0.001/0.000 = \infty$. With four decimal places we have $\hat{K}_m = 0.0010/0.0004 = 2.5$, a considerable improvement, but still about 6% from the precise answer, 2.359.

When fitting equations in several unknowns, the difficulties associated with rounding error become much worse. Most large modern computers carry out calculations with 12 or more significant digits, but even then serious

Table 10.1 ILLUSTRATION OF THE EFFECTS OF ROUNDING ERROR

For a sample set of five observations (in arbitrary units), equation 10.28 is used to calculate the value of $\hat{K}_m$. On the left-hand side, three decimal places are carried at each step in the calculation. On the right-hand side, the same calculation is carried out with four decimal places.

s	v	v^3	v^4	v^3/s	v^4/s	v^4/s^2	v^3	v^4	v^3/s	v^4/s	v^4/s^2
1	0.22	0.011	0.002	0.011	0.002	0.002	0.010 6	0.002 3	0.010 6	0.002 3	0.002 3
2	0.31	0.030	0.009	0.015	0.005	0.002	0.029 8	0.009 2	0.014 9	0.004 6	0.002 3
3	0.41	0.069	0.028	0.023	0.009	0.003	0.068 9	0.028 3	0.023 0	0.009 4	0.003 1
4	0.46	0.097	0.045	0.024	0.011	0.003	0.097 3	0.044 8	0.024 3	0.011 2	0.002 8
5	0.48	0.111	0.053	0.022	0.011	0.002	0.110 6	0.053 1	0.022 1	0.010 6	0.002 1
	Sums: 0.318	0.137	0.095	0.038	0.012		0.317 2	0.137 7	0.094 9	0.038 1	0.012 6

$$\hat{K}_m = \frac{0.137 \times 0.095 - 0.038 \times 0.318}{0.012 \times 0.318 - 0.038 \times 0.095}$$

$$= \frac{0.013 - 0.012}{0.004 - 0.004} = \frac{0.001}{0.000} = \infty$$

$$\hat{K}_m = \frac{0.137\,7 \times 0.094\,9 - 0.038\,1 \times 0.317\,2}{0.012\,6 \times 0.317\,2 - 0.038\,1 \times 0.094\,9}$$

$$= \frac{0.013\,1 - 0.012\,1}{0.004\,0 - 0.003\,6} = \frac{0.001\,0}{0.000\,4} = 2.50$$

difficulties can arise in the solution of simultaneous equations if one uses a badly designed program. One simple way of reducing (but not eliminating) trouble is to express all data in units chosen so that the numerical values are as close to 1.0 as possible, preferably between 0.1 and 10. It is far simpler to convert back into other units at the end of a computation than it is to search through a program to discover where an unexpected zero appeared.

A related problem arises when the simultaneous equations that one has to solve are either *singular* or *ill-conditioned*. Equations are said to be singular if they purport to contain more information than they in fact do. For example, the following pair of equations are singular:

$$x + y = 2$$
$$2x + 2y = 4$$

because the second equation contains exactly the same information as the first. Singular equations are always impossible to solve, and they always arise in regression problems if one attempts to fit equations that contain more parameters than the number of observations; for example, it would be hopeless to try to fit equation 10.45 to only two observations. This problem is readily avoided by the use of common sense; but ill-conditioned equations are more common and more difficult to avoid. This means that the simultaneous equations, although not strictly singular, are very nearly so, and require very precise computation for the unknowns to be evaluated, as with the following pair of equations:

$$x + y = 2$$
$$x + 1.000\,01y = 2.000\,01$$

In this case, the equations have a unique solution, $x = 1$, $y = 1$, but an increase of 0.000 01 in the right-hand side of the second equation would change the solution to $x = 0$, $y = 2$.

187

Ill-conditioning arises in regression problems if the observations contain little or no information about some of the parameters. For example, competitive inhibitors have most effect at low substrate concentrations, and a set of observations made exclusively at high substrate concentrations, however numerous, will often not permit evaluation of the inhibition constant. In more complex problems, ill-conditioning is often diagnosed by noticing that one parameter appears to have a relatively much higher standard error than the others. For example, in a case of mixed inhibition, if the standard errors of V, K_m and K_i' were all calculated to be about 5% of the parameter values, but that of K_i was about 50%, this would indicate that the experiment was designed badly for providing information about K_i and that more observations at low substrate concentrations were required. This is one of the occasions in which standard errors can be of real value, because although the numbers themselves may mean very little, comparisons between them may be very helpful.

It is not always certain that one is fitting data to the correct equation. In complex cases, such as with many of the models discussed in Chapter 7, the problem is often insoluble at the present state of experimental technique, because the expected differences between models are often less than experimental error. In these cases, one must admit defeat, because no statistical analysis can extract information that is not present in the data. However, in other cases there are several ways of recognizing failure to use the right equation, or *lack of fit*. The simplest and quickest way is to examine the signs of the residual deviations. If the correct equation is fitted there should be no discernible pattern; but if the first ten of twenty deviations were all negative, and the last ten were all positive, this would suggest lack of fit, as it is an improbable result *a priori*. Exactly this sort of result would arise if one fitted points from a sigmoid curve to the Michaelis–Menten equation.

A more precise test is possible if there are some observations in duplicate. These provide an independent measure of the experimental error, which may be compared with the estimate derived from the sum of squares, i.e. σ_{exp}^2. Provided that there are a reasonable number of duplicates (e.g. about ten in an experiment of about thirty observations; there is no need to carry out all observations in duplicate, and indeed it is better not to do so if this appreciably reduces the scope of the experiment) the estimates from the sum of squares and from the duplicates should be of the same order of magnitude. If they are not, and the estimate from the duplicates is much smaller, lack of fit is likely. The comparison can be made numerically, and tested with statistical tables (*see* Draper and Smith, 1966, pp. 26–32) but common sense is quicker and about as effective. It is often safer to avoid statistical tests unless one thoroughly understands their theoretical basis: if one blindly applies a textbook recipe and reaches a conclusion that is obviously wrong, one is in danger of looking foolish.

When lack of fit is diagnosed, it is necessary to find an equation that fits better. It is not sufficient to find one that gives a smaller sum of squares, as the introduction of more parameters into an equation invariably reduces the sum of squares. At the very least, the experimental variance, $\sigma_{exp}^2 = SS/(n-p)$, should be reduced and the residual deviations should no longer display symptoms of lack of fit. Again, statistical tests are available, but common sense is better.

10.9 Statistical aspects of the direct linear plot

The least-squares approach to regression problems has been presented in this chapter as the most convenient general method, rather than as the best. The reason for this is that to demonstrate that a least-squares solution to a problem is the 'best' solution, one must assume (*i*) that the random errors in the measurements are normally distributed; (*ii*) that only one measured variable is subject to experimental error; (*iii*) that the correct weights are known; and (*iv*) that systematic error can be ignored. Unfortunately, one knows very little about the truth of any of these assumptions in practice. Most scientists prefer their conclusions to depend on as few unproved assumptions as possible, and in the last half-century a separate branch of statistics has developed, known as *non-parametric* or *distribution-free* statistics, which is predicated on minimal assumptions. Of the assumptions listed above, the first three are dropped but the last must be retained: indeed, it is clearly impossible to allow for systematic errors that have eluded all efforts at detection. This last assumption is usually expressed in the simpler form that, in the absence of other information, we assume that the error in any measurement is as likely to be positive as to be negative.

The simplest example of a non-parametric statistic is the median as a measure of the 'average' of a sample rather than the mean. In order to find the median of a set of numbers, we first arrange them in order from the lowest to the highest value and then take the middle value; if there are an even number of values altogether, we take the mean of the middle two values. The great practical merit of the median is that it is almost unaffected by the presence of a few very bad values, or *outliers*. The mean, on the other hand, is very seriously affected by outliers. The disadvantage of the median is that, if the observations are truly normally distributed, then the median is less 'efficient' than the mean, in the sense that the standard deviation of the median of a sample is, at worst, about 25% greater than that of the mean. For small numbers of observations, the difference in efficiency between the mean and the median is less than this value. More important, if the distribution is not normal, but contains even a small proportion of outliers (e.g. if 10% of values have a standard deviation three times greater than that of the other 90%), the median becomes more efficient than the mean. Another major advantage of the median is that it does not require weighting in order to provide a good estimate, because the best values in a set tend to be found near the middle of the range, and the worst near the extremes. Thus, although it is possible to weight a median, there is little advantage in doing so. In contrast, it is essential to weight a mean if it is to provide a good estimate.

The direct linear plot of Eisenthal and Cornish-Bowden (1974) (*see* Section 2.5) represents an attempt at introducing non-parametric ideas into the estimation of enzyme kinetic parameters, at the same time greatly simplifying the procedures and concepts. For any non-duplicate pair of observations

189

(s_i, v_i) and (s_j, v_j), there is a unique pair of values of the Michaelis–Menten equation (K_{ij}, V_{ij}) that satisfy both observations exactly, given by

$$K_{ij} = \frac{v_j - v_i}{\dfrac{v_i}{s_i} - \dfrac{v_j}{s_j}}$$

$$V_{ij} = \frac{s_i - s_j}{\dfrac{s_i}{v_i} - \dfrac{s_j}{v_j}}$$

These values are defined by the coordinates of the point of intersection of the lines drawn for the two observations as described in Section 2.5. Altogether n observations provide $\frac{1}{2}n(n-1)$ such pairs of values (fewer if there are duplicates). The median of the set of K_{ij} can be defined as $\hat{K}_m$, and the median of the set of V_{ij} as $\hat{V}$. These median values can be found very simply from the direct linear plot, as indicated in *Figure 10.3*. In order to test the validity of

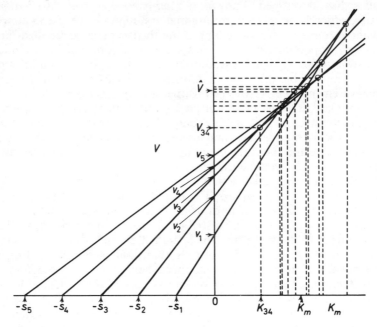

Figure 10.3 Determination of median estimates from the direct linear plot: The lines are drawn as in Figure 2.7, and each intersection (shown as a circle) provides an estimate K_{ij} of K_m, and an estimate V_{ij} of V. These estimates are marked off on the axes for clarity. $\hat{K}_m$ is then taken as the median K_{ij} value, and $\hat{V}$ as the median V_{ij} value. If there are an even number of intersections, as in this example, the median is taken as the mean of the middle two values

this procedure, Cornish-Bowden and Eisenthal (1974) carried out a computer simulation of many thousands of experiments, incorporating various different assumptions about the nature of experimental error. Computer-simulated experiments have several advantages over real experiments in this type of study: it is possible to carry out far more of them; the true values of

190

the parameters are known; and the true distribution of experimental error is known. They found that the least-squares estimates were better than the median estimates in experiments where all of the assumptions embodied in the least-squares approach were correct. This result was, of course, expected, but the difference was surprisingly slight, and the median estimates came closer to the true values in about 40% of experiments. However, in experiments where the least-squares assumptions were not true, e.g. the data contained outliers, or the weighting scheme was incorrect, or there were errors in s as well as v, the slight advantage of the least-squares estimates disappeared. Cornish-Bowden and Eisenthal concluded that, with realistic assumptions about experimental error, there was no reason to prefer the least-squares method.

The direct linear plot also provides a simple means of defining joint confidence limits for $\mathscr{K}_m$ and $\mathscr{V}$. At the simplest level, the scattering of the intersection points provides a clear qualitative picture of the precision of the parameters. A more exact result is obtained from consideration of the fact that each region bounded by the lines of the plot corresponds to a different permutation of signs among the residual deviations, as indicated in *Figure 10.4*. This is because each line defines the boundary between the set of

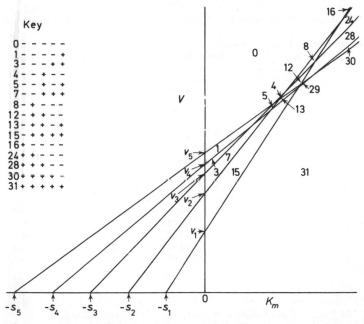

Figure 10.4 Permutations of signs among the deviations: This plot shows the same data as Figure 10.3, but re-labelled so as to indicate that each region between the lines corresponds to a different permutation of signs among the residual deviations. For example, if the first deviation is negative and the other four are positive, the values of $\mathscr{K}_m$ and $\mathscr{V}$ must lie within the region labelled 15. The labels are obtained by treating each − as a binary zero and each + as a binary unit, and converting the resulting binary number (01111 for region 15) into its decimal equivalent. A conversion table is given as an inset in the figure. Certain conceivable regions, e.g. region 2 (for − − − + −), are missing from the plot: these correspond to permutations of signs that are impossible for the data shown, although they are not impossible in general

191

(K_m, V) pairs for which the corresponding deviation is positive and the set of (K_m, V) pairs for which it is negative. As, from the fundamental assumption of non-parametric statistics, all permutations of signs for the true errors are equally likely *a priori*, it is fairly easy to find joint confidence limits for $\mathcal{K}_m$ and $\mathcal{V}$, as will now be shown.

If the total number of positive signs is counted, one can obtain open-ended confidence regions, which are rigorous but inconvenient as they fail to exclude some obviously wrong estimates of the parameters. The reason for this is that it is easy to draw an obviously ill-fitting line through a set of points that nonetheless passes half of the points on one side and half on the other.

More satisfactory confidence limits are found by searching for estimates of $\mathcal{K}_m$ and $\mathcal{V}$ that as nearly as possible predict alternate positive and negative errors. In other words, we try to maximize the number of 'runs' of signs among the deviations. On the direct linear plot, this approach favours the small enclosed regions close to the median estimates over the infinite open regions at the edges. If one takes as a confidence region the jagged region formed by combining all of the enclosed regions, the actual level of confidence depends on the number of observations, as indicated in *Table 10.2*. If there are twelve

Table 10.2 PROBABILITY (EXPRESSED IN %) OF AT LEAST m 'RUNS' AMONG n RANDOM SIGNS

A region with $m \geqslant 3$ can be found on the direct linear plot by taking all regions completely enclosed by the lines of the plot. A region with $m \geqslant 4$ is most easily found by labelling the plot as indicated in *Figure 10.4*.

n	m ⩾ 3	m ⩾ 4
3	25.0	0.0
4	50.0	12.5
5	68.8	31.3
6	81.3	50.0
7	89.1	65.6
8	93.8	77.3
9	96.5	85.5
10	98.0	91.0
11	98.9	94.5
12	99.41	96.7
13	99.68	98.1
14	99.83	98.9
15	99.91	99.35

or more observations, one can find a smaller region of 95% confidence or better. More detail about this method and its principle is given by Cornish-Bowden and Eisenthal (1974). For most purposes, a qualitative interpretation of the plot is likely to suffice.

The principal advantages of the direct linear plot as a method of evaluating $\hat{K}_m$ and $\hat{V}$ are that it is very simple to use, it is insensitive to isolated bad observations and it does not assume accurate knowledge of the relative precision of each observation. It is unfortunately impractical to extend the method to the fitting of more complex equations of three or more parameters, because

192

the computational labour increases very steeply with the number of parameters and with the number of observations. One can avoid this difficulty, however, by the secondary-plot approach described in earlier chapters for conventional plotting procedures. For example, in a two-substrate experiment, one can use the direct linear plot as a primary plot to determine $\hat{K}_m^{app}$, $\hat{V}^{app}$ and $\hat{V}^{app}/\hat{K}_m^{app}$ at various constant concentrations of the second substrate. One can then take advantage of the fact that the expressions for V^{app} and V^{app}/K_m^{app} are often of the same form as the Michaelis–Menten equation, e.g. (equation 5.6) $V^{app} = Vb/(K_m^B + b)$, and use secondary plots (of V against K_m^B in this case, with intercepts V^{app} on the V axis and $-b$ on the K_m^B axis) to determine the four parameters.

10.10 Final note

This chapter is inevitably a compromise between the much shorter (or absent) account that one generally finds in a biochemistry book and the complete book that could profitably be written. It has been necessary to omit much of the background statistical knowledge that is useful in the practice of regression. This is particularly unfortunate in view of the dearth of good statistics books at an appropriate level: statistics books are divided between those that are rigorous but largely incomprehensible and those that contain a short last chapter on straight-line fitting at the end of a procession of recipes for t-tests, F-tests and the other paraphernalia of classical statistics. Fortunately, there does exist one book, 'Lectures in Biostatistics' (Colquhoun, 1971), that combines readability, honesty (about how little we usually know about the validity of our assumptions) and relevance to real problems, qualities that are sadly lacking from most other textbooks on the subject.

References

The sections of the book where the work is referred to are given in brackets after each reference.

G. S. ADAIR (1925a) *J. biol. Chem.* **63**, 529–545. [7.4]

G. S. ADAIR (1925b) *Proc. Roy. Soc., Ser. A* **109**, 292–300. [7.4]

R. A. ALBERTY (1953) *J. Amer. chem. Soc.* **75**, 1928–1932. [5.4]

R. A. ALBERTY (1958) *J. Amer. chem. Soc.* **80**, 1777–1782. [5.4]

R. A. ALBERTY and B. M. KOERBER (1957) *J. Amer. chem. Soc.* **79**, 6379–6382. [8.6]

R. A. ALBERTY, V. MASSEY, C. FRIEDEN and A. R. FUHLBRIGGE (1954) *J. Amer. chem. Soc.* **76**, 2485–2493. [5.8]

S. ARRHENIUS (1889) *Z. physik. Chem.* **4**, 226–248. [1.6]

R. E. BARDEN, C.-H. FUNG, M. F. UTTER and M. C. SCRUTTON (1972) *J. biol. Chem.* **247**, 1323-1333. [5.2]

M. L. BENDER, M. L. BEGUÉ-CANTÓN, R. L. BLAKELEY, L. J. BRUBACHER, J. FEDER, C. R. GUNTER, F. J. KÉZDY, J. V. KILLHEFFER, JR., T. H. MARSHALL, C. G. MILLER, R. W. ROESKE and J. K. STOOPS (1966) *J. Amer. chem. Soc.* **88**, 5890–5913. [9.3]

M. L. BENDER, F. J. KÉZDY and C. R. GUNTER (1964) *J. Amer. chem. Soc.* **86**, 3714–3721. [6.8]

D. BLANGY, H. BUC and J. MONOD (1968) *J. mol. Biol.* **31**, 13–35. [7.7]

V. BLOOMFIELD, L. PELLER and R. A. ALBERTY (1962) *J. Amer. chem. Soc.* **84**, 4367–4374. [5.4]

R. M. BOCK and R. A. ALBERTY (1953) *J. Amer. chem. Soc.* **75**, 1921–1925. [2.6]

M. BODENSTEIN (1913) *Z. physik. Chem.* **85**, 329–397. [2.4]

C. BOHR (1903) *Zentralbl. Physiol.* **17**, 682–688. [7.2]

J. BOTTS and M. MORALES (1953) *Trans. Faraday Soc.* **49**, 696–707. [3.4, 3.7, 4.1, 4.7]

P. D. BOYER (1959) *Arch. Biochem. Biophys.* **82**, 387–410. [5.9]

G. E. BRIGGS and J. B. S. HALDANE (1925) *Biochem. J.* **19**, 338–339. [2.3, 2.4]

H. G. BRITTON (1966) *Arch. Biochem. Biophys.* **117**, 167–183. [5.10]

H. G. BRITTON (1973) *Biochem. J.* **133**, 255–261. [5.10]

H. G. BRITTON and J. B. CLARKE (1968) *Biochem. J.* **110**, 161–179. [5.10]

A. J. BROWN (1892) *J. chem. Soc. (Trans.)* **61**, 369–385. [2.1]

A. J. BROWN (1902) *J. chem. Soc. (Trans.)* **81**, 373–388. [2.1]

E. BUCHNER (1897) *Ber. dt. chem. Ges.* **30**, 117–124. [2.1]

J. J. BURKE, G. G. HAMMES and T. B. LEWIS (1965) *J. chem. Phys.* **42**, 3520–3525. [9.6]

H. CEDAR and J. H. SCHWARTZ (1969) *J. biol. Chem.* **244**, 4122–4127. [5.9]

S. CHA (1968) *J. biol. Chem.* **243**, 820–825. [3.6]

B. CHANCE (1943) *J. biol. Chem.* **151**, 553–557. [2.3]

B. CHANCE (1963) in S. L. Friess, E. S. Lewis and A. Weissberger (Editors), *Technique of Organic Chemistry* 2nd edn, Interscience, New York, Vol. 8, Part 2, pp. 1314–1360. [5.4]

W. W. CLELAND (1963) *Biochim. Biophys. Acta* **67**, 104–137. [5.3, 5.4]

194

W. W. CLELAND (1967) *Advan. Enzymol.* **29**, 1–32. [10.5]

D. COLQUHOUN (1971) *Lectures in Biostatistics* Clarendon Press, Oxford. [10.6, 10.10]

A. CONWAY and D. E. KOSHLAND, JR. (1968) *Biochemistry* **7**, 4011–4023. [7.8]

A. J. CORNISH-BOWDEN (1972) *Biochem. J.* **130**, 637–639. [8.7]

A. CORNISH-BOWDEN (1974) *Biochem. J.* **137**, 143–144. [4.5]

A. CORNISH-BOWDEN (1975) *Biochem. J.* **149**, 305–312. [8.5]

A. CORNISH-BOWDEN and R. EISENTHAL (1974) *Biochem. J.* **139**, 721–730. [10.9]

A. CORNISH-BOWDEN and D. E. KOSHLAND, JR. (1970) *J. biol. Chem.* **245**, 6241–6250. [7.9]

A. J. CORNISH-BOWDEN and D. E. KOSHLAND, JR. (1971) *J. biol. Chem.* **246**, 3092–3102. [7.9]

K. DALZIEL (1957) *Acta Chem. Scand.* **11**, 1706–1723. [5.4]

K. DALZIEL (1962) *Nature, Lond.* **196**, 1203–1205. [2.3]

F. DE MIGUEL MERINO (1974) *Biochem. J.* **143**, 93–95. [Appendix 2.1]

H. B. F. DIXON (1973) *Biochem. J.* **131**, 149–154. [6.2]

M. DIXON (1953) *Biochem. J.* **55**, 170–171. [4.5]

M. DOUDOROFF, H. A. BARKER and W. Z. HASSID (1947) *J. biol. Chem.* **168**, 725–732. [5.2, 5.9]

N. R. DRAPER and H. SMITH (1966) *Applied Regression Analysis* Wiley, New York. [10.7, 10.8]

G. S. EADIE (1942) *J. biol. Chem.* **146**, 85–93. [2.5]

M. EIGEN (1954) *Discuss. Faraday Soc.* **17**, 194–205. [9.5]

M. EIGEN and G. G. HAMMES (1963) *Advan. Enzymol.* **25**, 1–38. [2.3]

R. EISENTHAL and A. CORNISH-BOWDEN (1974) *Biochem. J.* **139**, 715–720. [2.5, 10.9]

W. D. ELLIS and H. B. DUNFORD (1968) *Biochemistry* **7**, 2054–2062. [9.4]

H. VON EULER and K. JOSEPHSON (1924) *Hoppe-Seyler's Z. physiol. Chem.* **136**, 30–44. [2.6]

H. EYRING (1935) *J. chem. Phys.* **3**, 107–115. [1.7]

W. FERDINAND (1966) *Biochem. J.* **98**, 278–283. [7.11]

E. FISCHER (1894) *Ber. dt. chem. Ges.* **27**, 2985–2993. [7.6]

C. FRIEDEN (1967) *J. biol. Chem.* **242**, 4045–4052. [7.10]

C. FRIEDEN and R. F. COLMAN (1967) *J. biol. Chem.* **242**, 1705–1715. [7.10]

H. J. FROMM (1970) *Biochem. Biophys. Res. Commun.* **40**, 692–697. [3.4]

N. K. GHOSH and W. H. FISHMAN (1966) *J. biol. Chem.* **241**, 2516–2522. [4.4]

E. A. GUGGENHEIM (1926) *Phil. Mag., Ser. VII* **2**, 538–543. [1.5]

J. S. GULBINSKY and W. W. CLELAND (1968) *Biochemistry* **7**, 566–575. [3.6, 5.2, 7.11]

H. GUTFREUND (1955) *Discuss. Faraday Soc.* **20**, 167–173. [9.2]

J. E. HABER and D. E. KOSHLAND, JR. (1967) *Proc. natn. Acad. Sci. U.S.* **58**, 2087–2093. [7.8]

J. B. S. HALDANE (1930) *Enzymes* Longmans Green, London. [2.6, 5.2, 6.4]

G. G. HAMMES (1968) *Advan. Protein Chem.* **23**, 1–57. [9.5]

G. G. HAMMES and P. FASELLA (1962) *J. Amer. chem. Soc.* **84**, 4644–4650. [9.5]

G. G. HAMMES and P. FASELLA (1963) *J. Amer. chem. Soc.* **85**, 3929–3932. [9.5]

G. G. HAMMES and C. N. PACE (1968) *J. phys. Chem.* **72**, 2227–2230. [9.6]

G. G. HAMMES and P. R. SCHIMMEL (1970) in P. Boyer (Editor), *The Enzymes* 3rd edn, Academic Press, New York, Vol. 2, pp. 67–114. [9.5]

C. S. HANES (1932) *Biochem. J.* **26**, 1406–1421. [2.5]

A. V. HARCOURT (1867) *J. chem. Soc.* **20**, 460–492. [1.6]

B. S. HARTLEY and B. A. KILBY (1954) *Biochem. J.* **56**, 288–297. [9.3]

V. HENRI (1902) *C. r. hebd. Acad. Sci., Paris* **135**, 916–919. [2.1]

V. HENRI (1903) *Lois Générales de l'Action des Diastases* Hermann, Paris. [2.1]

A. V. HILL (1910) *J. Physiol.* **40**, iv–vii. [7.2, 7.3]

D. I. HITCHCOCK (1926) *J. Amer. chem. Soc.* **48**, 2870. [2.2]

B. H. J. HOFSTEE (1952) *J. biol. Chem.* **199**, 357–364. [2.5]

B. H. J. HOFSTEE (1959) *Nature, Lond.* **184**, 1296–1298 (with additional comments by M. Dixon and E. C. Webb). [2.5]

R. Y. HSU, W. W. CLELAND and L. ANDERSON (1966) *Biochemistry* **5**, 799–807. [5.7]

H. T. HUANG and C. NIEMANN (1951) *J. Amer. chem. Soc.* **73**, 1541–1548. [8.3]

A. HUNTER and C. E. DOWNS (1945) *J. biol. Chem.* **157**, 427–446. [4.3, 4.4]

J. C. HWANG, C. H. CHEN and R. H. BURRIS (1973) *Biochim. Biophys. Acta* **292**, 256–270. [10.6]

D. W. INGLES and J. R. KNOWLES (1967) *Biochem. J.* **104**, 369–377. [4.8, 6.8]

INTERNATIONAL UNION OF BIOCHEMISTRY (1961) *Report of the Commission on Enzymes* Pergamon Press, Oxford. [5.4]

R. R. JENNINGS and C. NIEMANN (1955) *J. Amer. chem. Soc.* **77**, 5432–5433. [8.3]

G. JOHANSEN and R. LUMRY (1961) *C. r. trav. Lab. Carlsberg* **32**, 185–214. [2.5, 10.4]

J. C. KENDREW, R. E. DICKERSON, B. E. STRANDBERG, R. G. HART, D. R. DAVIES, D. C. PHILLIPS and V. C. SHORE (1960) *Nature, Lond.* **185**, 422–427. [7.2]

F. J. KÉZDY, J. JAZ and A. BRUYLANTS (1958) *Bull. Soc. chim. Belg.* **67**, 687–706. [1.5]

E. L. KING and C. ALTMAN (1956) *J. phys. Chem.* **60**, 1375–1378. [3.1–7, 6.4]

M. E. KIRTLEY and D. E. KOSHLAND, JR. (1967) *J. biol. Chem.* **242**, 4192–4205. [7.8]

J. R. KNOWLES (1965) *Biochem. J.* **95**, 180–190. [8.6]

J. R. KNOWLES, H. SHARP and P. GREENWELL (1969) *Biochem. J.* **113**, 343–351. [2.3]

D. E. KOSHLAND, JR. (1954) in W. D. McElroy and B. Glass (Editors), *A Symposium on the Mechanism of Enzyme Action* Johns Hopkins Press, Baltimore, pp. 608–641. [5.2]

D. E. KOSHLAND, JR. (1955) *Discuss. Faraday Soc.* **20**, 142–148. [5.9]

D. E. KOSHLAND, JR. (1958) *Proc. natn. Acad. Sci. U.S.* **44**, 98–99. [5.2, 7.6]

D. E. KOSHLAND, JR. (1959a) in P. D. Boyer, H. Lardy and K. Myrbäck (Editors), *The Enzymes* 2nd edn, Academic Press, New York, Vol. 1, pp. 305–346. [5.2, 7.6]

D. E. KOSHLAND, JR. (1959b) *J. cell. comp. Physiol.* **54**, suppl., 245–258. [5.2, 7.6]

D. E. KOSHLAND, JR., G. NÉMETHY and D. FILMER (1966) *Biochemistry* **5**, 365–385. [7.8]

D. E. KOSHLAND, JR., D. H. STRUMEYER and W. J. RAY, JR. (1962) *Brookhaven Symp. Biol.* **15**, 101–133. [9.3]

K. J. LAIDLER (1955) *Can. J. Chem.* **33**, 1614–1624. [2.4]

K. J. LAIDLER (1965) *Chemical Kinetics* 2nd edn, McGraw Hill, New York, Ch. 3. [1.7]

I. LANGMUIR (1916) *J. Amer. chem. Soc.* **38**, 2221–2295. [2.2]

I. LANGMUIR (1918) *J. Amer. chem. Soc.* **40**, 1361–1403. [2.2, 2.5]

A. LEVITZKI and D. E. KOSHLAND, JR. (1969) *Proc. natn. Acad. Sci. U.S.* **62**, 1121–1128. [7.9]

A. LEVITZKI, W. B. STALLCUP and D. E. KOSHLAND, JR. (1971) *Biochemistry* **10**, 3371–3378. [7.9]

H. LINEWEAVER and D. BURK (1934) *J. Amer. chem. Soc.* **56**, 658–666. [2.2, 2.5]

R. A. MACQUARRIE and S. A. BERNHARD (1971) *J. mol. Biol.* **55**, 181–192. [7.9]

H. R. MAHLER and E. H. CORDES (1966) *Biological Chemistry* Harper and Row, New York, pp. 219–277. [5.4]

R. G. MARTIN (1963) *J. biol. Chem.* **238**, 257–268. [7.7]

S. J. MASON (1953) *Proc. Inst. Radio Engrs* **41**, 1144–1156. [3.4]

S. J. MASON (1956) *Proc. Inst. Radio Engrs* **44**, 920–926. [3.4]

L. MICHAELIS (1926) *Hydrogen Ion Concentration* translated from 2nd German edn (1921) by W. A. Perlzweig, Baillière, Tindall and Cox, London, Vol. 1. [6.2]

L. MICHAELIS and M. L. MENTEN (1913) *Biochem. Z.* **49**, 333–369. [2.2, 2.5, 8.1]

L. MICHAELIS and H. PECHSTEIN (1914) *Biochem. Z.* **60**, 79–90. [2.7]

J. MONOD, J.-P. CHANGEUX and F. JACOB (1963) *J. mol. Biol.* **6**, 306–329. [7.7]

J. MONOD, J. WYMAN and J.-P. CHANGEUX (1965) *J. mol. Biol.* **12**, 88–118. [3.8, 7.2, 7.7]

D. K. MYERS (1952) *Biochem. J.* **51**, 303–311. [4.10]

J. M. NELSON and R. S. ANDERSON (1926) *J. biol. Chem.* **69**, 443–448. [4.3]

E. A. NEWSHOLME and W. GEVERS (1967) *Vitam. Horm.* **25**, 1–87. [7.11]

L. W. NICHOL, W. J. H. JACKSON and D. J. WINZOR (1967) *Biochemistry* **6**, 2449–2456. [7.10]

D. B. NORTHROP (1969) *J. biol. Chem.* **244**, 5808–5819. [5.2]

C. O'SULLIVAN and F. W. TOMPSON (1890) *J. chem. Soc. (Trans.)* **57**, 834–931. [2.1]

L. OUELLET and J. A. STEWART (1959) *Can. J. Chem.* **37**, 737–743. [9.3]

L. PAULING (1935) *Proc. natn. Acad. Sci. U.S.* **21**, 186–191. [7.5]

M. F. PERUTZ, M. G. ROSSMANN, A. F. CULLIS, H. MUIRHEAD, G. WILL and A. C. T. NORTH (1960) *Nature, Lond.* **185**, 416–422. [7.2, 7.5]

B. R. RABIN (1967) *Biochem. J.* **102**, 22c–23c. [7.11]

W. J. RAY, JR. and G. A. ROSCELLI (1964) *J. biol. Chem.* **239**, 3935–3941. [5.10]

J. G. REICH (1970) *FEBS Lett.* **9**, 245–251. [10.1]

F. SCHØNHEYDER (1952) *Biochem. J.* **50**, 378–384. [8.3]

G. R. SCHONBAUM, B. ZERNER and M. L. BENDER (1961) *J. biol. Chem.* **236**, 2930–2935. [9.3]

G. W. SCHWERT (1969) *J. biol. Chem.* **244**, 1278–1284. [8.6]

H. L. SEGAL, J. F. KACHMAR and P. D. BOYER (1952) *Enzymologia* **15**, 187–198. [5.4]

F. SEYDOUX, O. P. MALHOTRA and S. A. BERNHARD (1974) *Crit. Rev. Biochem.* **2**, 227–257. [7.9]

S. P. L. SØRENSEN (1909) *C. r. trav. Lab. Carlsberg* **8**, 1–168 (in French; a German version appeared in *Biochem. Z.* **21**, 131–304). [2.2, 6.1]

P. A. SRERE (1968) in T. W. Goodwin (Editor), *The Metabolic Roles of Citrate* Academic Press, London, pp. 11–21. [4.10]

W. B. STALLCUP and D. E. KOSHLAND, JR. (1973) *J. mol. Biol.* **80**, 77–91. [7.9]

O. H. STRAUS and A. GOLDSTEIN (1943) *J. gen. Physiol.* **26**, 559–585. [4.10]

W. H. SWANN (1969) *FEBS Lett.* **2**, *suppl.*, 39–55. [10.7]

J. R. SWEENY and J. R. FISHER (1968) *Biochemistry* **7**, 561–565. [7.11]

E. S. SWINBOURNE (1960) *J. chem. Soc.* 2371–2372. [1.5]

K. TAKETA and B. M. POGELL (1965) *J. biol. Chem.* **240**, 651–662. [7.3]

H. THEORELL and B. CHANCE (1951) *Acta Chem. Scand.* **5**, 1127–1144. [5.2]

D. D. VAN SLYKE and G. E. CULLEN (1914) *J. biol. Chem.* **19**, 141–180. [2.2, 2.4]

J. H. VAN'T HOFF (1884) *Études de Dynamique Chimique* Muller, Amsterdam, pp. 114–118. [1.6]

R. O. VIALE (1970) *J. theor. Biol.* **27**, 377–385. [2.1]

M. V. VOLKENSTEIN and B. N. GOLDSTEIN (1966) *Biochim. Biophys. Acta* **115**, 471–477. [3.4–5]

A. C. WALKER and C. L. A. SCHMIDT (1944) *Arch. Biochem.* **5**, 445–467. [8.2]

C. W. WHARTON, A. CORNISH-BOWDEN, K. BROCKLEHURST and E. M. CROOK (1974) *Biochem. J.* **141**, 365–381. [10.7]

G. N. WILKINSON (1961) *Biochem. J.* **80**, 324–332. [2.5, 10.4]

J. T.-F. WONG and C. S. HANES (1962) *Can. J. Biochem. Physiol.* **40**, 763–804. [5.2].

B. WOOLF (1929) *Biochem. J.* **23**, 472–482. [5.2]

B. WOOLF (1931) *Biochem. J.* **25**, 342–348. [5.2]

B. WOOLF (1932), cited by J. B. S. Haldane and K. G. Stern in *Allgemeine Chemie der Enzyme* Steinkopff, Dresden and Leipzig, pp. 119–120. [2.5]

C. C. WRATTEN and W. W. CLELAND (1965) *Biochemistry* **4**, 2442–2451. [4.7]

A. WURTZ (1880) *C. r. hebd. Acad. Sci., Paris* **91**, 787–791. [2.1]

Index